Kawasaki 650 Four Owners Workshop Manual

by Pete Shoemark

Models covered
Z650 B1. 652cc. UK October 1976 on
KZ650 B1 652cc. US September 1976 on

ISBN 978 0 85696 373 5

(373-8Q4)

J H Haynes & Co. Ltd.
Haynes North America, Inc

www.haynes.com

Disclaimer

There are risks associated with automotive repairs. The ability to make repairs depends on the individual's skill, experience and proper tools. Individuals should act with due care and acknowledge and assume the risk of performing automotive repairs.

The purpose of this manual is to provide comprehensive, useful and accessible automotive repair information, to help you get the best value from your vehicle. However, this manual is not a substitute for a professional certified technician or mechanic.

This repair manual is produced by a third party and is not associated with an individual vehicle manufacturer. If there is any doubt or discrepancy between this manual and the owner's manual or the factory service manual, please refer to the factory service manual or seek assistance from a professional certified technician or mechanic.

Even though we have prepared this manual with extreme care and every attempt is made to ensure that the information in this manual is correct, neither the publisher nor the author can accept responsibility for loss, damage or injury caused by any errors in, or omissions from, the information given.

Acknowledgements

Our thanks are due to Kawasaki Motors (UK) Ltd, who supplied the test machine used in the photographic sequences. We would like to thank in particular Nick Jeffery and Clive Stokes of the above Company, for their assistance with the project.

Gregory S. Owens of Newbury Park, California kindly supplied the Kawasaki KZ650 2 shown on the front cover of the manual.

Brian Horsfall assisted with the stripdown and rebuilding, and devised the ingenious methods for overcoming the lack of service tools. Les Brazier arranged and took the cover and inside cover photographs, also the photographs which accompany the text.

Jeff Clew edited the text and compiled the source of technical data used in the origination of the manual.

Finally, we would also like to thank the Avon Rubber Company, who supplied information and technical assistance on tyre fitting; NGK Spark Plugs (UK) Ltd for information on spark plug maintenance and electrode conditions, and Renold Ltd for advice on chain care and renewal.

About this manual

The author of this manual has the conviction that the only way in which a meaningful and easy to follow text can be written is to first do the work himself, under conditions similar to those found in the average home. As a result, the hands seen in the photographs are the hands of the author or those of another engineer who assisted. The machine photographed was a used model that had covered three thousand miles, so that the conditions encountered would be similar to those found by the average rider.

Unless specially mentioned, and therefore considered essential, Kawasaki service tools have not been used. There is invariably some alternative means of slackening or removing some vital component when service tools are not available and the risk of damage has to be avoided at all costs.

Each of the six Chapters is divided into numbered Sections. Within the Sections are numbered paragraphs. In consequence, cross reference throughout this manual is both straightforward and logical. When a reference is made 'See Section 1.6', it means see Section 1, paragraph 6 in the same Chapter. If another Chapter were meant, the text would read 'See Chapter 4, Section 1.6'. All the photographs are captioned with a Section paragraph number to which they refer and are always relevant to the Chapter text adjacent.

Figure numbers (usually line illustrations) appear in numerical order, within a given Chapter. Figure 1.2 therefore refers to the first figure in Chapter 1.

Left-hand and right-hand descriptions of parts of the machine or the machine itself, refer to the right and left side of the machine, with the rider seated in the normal riding position.

Motorcycle manufacturers continually make changes to specifications and recommendations, and these, when notified, are incorporated into our manuals at the earliest opportunity.

Contents

Note: General descriptions and specifications are given in each Chapter immediately after list of contents.
Fault diagnosis is given at the end of the appropriate Chapter.

Right-hand view of 1977 Kawasaki Z650

Left-hand view of 1977 Kawasaki Z650

Introduction to the Kawasaki 650 Four

When the 900 cc Z1 model was first introduced in 1972 it was obvious that Kawasaki had scored a huge success. The growth of the company had been little short of phenomenal, perhaps causing some people to wonder how it was achieved. The answer lay in the vast resources of the firm and the extent of their technological know-how, which extended into railroad, shipping, and aircraft transportation on a grand scale. All these activities rolled into one form, Kawasaki Heavy Industries, a giant manufacturing complex that produces an astonishing variety of products and markets them all over the world.

When the 650 cc four was introduced in 1976, it was no real surprise that it closely resembled the Z1 model, the latter having proven to be both fast and reliable. A comparison of the two machines will show just how close this resemblance is.

In just a few years Kawasaki have become the fourth largest motorcycle manufacturer in the world and that in itself is quite an accomplishment when it is recalled that some European companies have been manufacturing machines for over 60 years. Kawasaki have now become seriously involved with racing, to such an extent that they field factory teams in trials, road racing and motocross, and were the only Japanese manufacturer to participate in the 48th ISDT held in Berkshire, USA. More important is the readiness of Kawasaki to incorporate the hard learned lessons, acquired at the race track, in their road going machines. In this way they have successfully capitalised on their competition successes, putting the knowledge they have gained at the disposal of all those who purchase their high quality products.

Dimensions and weight

Dimensions	European models	US models
Overall length	2220 mm (87·4 in)	2170 mm (85·43 in)
Overall width	850 mm (33·5 in)	
Overall height	1145 mm (45·1 in)	
Wheelbase	1420 mm (55·91 in)	
Ground clearance	140 mm (5·51 in)	145 mm (5·71 in)
Weight (dry)	211 kg (465·2 lbs)	

Ordering spare parts

When ordering spare parts for a Kawasaki it is advisable to deal direct with an official Kawasaki agent who should be able to supply most of the parts from stock. Parts cannot be obtained direct from Kawasaki UK; all orders must be routed via an approved agent as is common with most other makes.

Always quote the frame and engine numbers in full. The frame number is stamped on the left-hand side of the steering head and the engine number on top of the crankcase to the rear of the cylinder block, on the right-hand side.

It is always best to quote the colour scheme for any of the cycle parts that have to be ordered. Use only genuine Kawasaki parts. Pattern parts should be avoided as they are usually inferior in quality. Some of the more expendable parts such as bulbs, spark plugs, chains, tyres, oils and greases etc., can be obtained from accessory stores and motor factors, who have convenient opening hours and can often be found nearer home. It is also possible to obtain parts on a Mail Order basis from a number of specialists who advertise regularly in the motorcycle magazines.

Engine Number Location

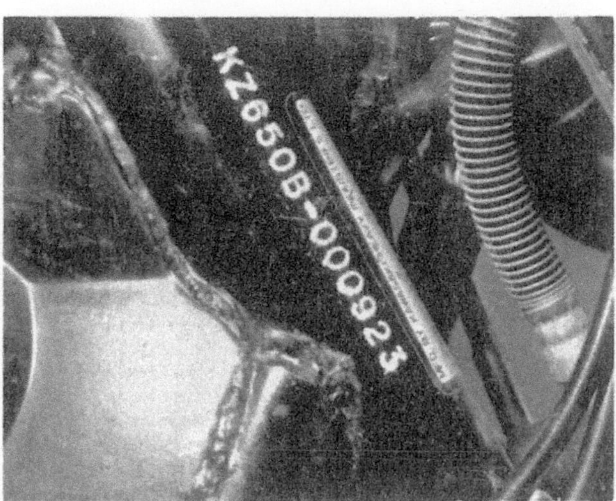

Frame Number Location

Routine maintenance

Introduction

Periodic routine maintenance is a continuous process that commences immediately the machine is used. It must be carried out at specified mileage recording, or on a calendar basis if the machine is not used frequently, whichever is the sooner. Maintenance should be regarded as an insurance policy, to help keep the machine in the peak of condition and to ensure long, trouble-free service. It has the additional benefit of giving early warning of any faults that may develop and will act as a regular safety check, to the obvious advantage of both rider and machine alike.

The various maintenance tasks are described under their respective mileage and calendar headings. Accompanying diagrams are provided, where necessary. It should be remembered that the interval between the various maintenance tasks serves only as a guide. As the machine gets older or is used under particularly adverse conditions, it would be advisable to reduce the period between each check.

For ease of reference each service operation is described in detail under the relevant heading. However, if further general information is required, it can be found within the manual under the pertinent section heading in the relevant Chapter.

In order that the routine maintenance tasks are carried out with as much ease as possible, it is essential that a good selection of general workshop tools is available.

Included in the kit must be a range of metric ring or combination spanners, a selection of crosshead screwdrivers and at least one pair of circlip pliers.

Additionally, owing to the extreme tightness of most casing screws on Japanese machines, an impact screwdriver, together with a choice of large or small crosshead screw bits, is absolutely indispensable. This is particuarly so if the engine has not been dismantled since leaving the factory.

Daily

A daily check of the motorcycle is essential both from mechanical and safety aspects. It is a good idea to develop this checking procedure in a specific sequence so that it will ultimately become as instinctive as actually riding the machine. Done properly, this simple checking sequence will give advanced warning of impending mechanical failures and any condition which may jeopardise the safety of the rider.

1 *Oil level*

The level of the engine oil is quickly checked by way of the oil sight glass set in the right-hand outer casing. With the machine standing on level ground, the oil should be visible half way up the plastic window. Marks are provided on the rim of the window, indicating the maximum and minimum oil levels. If necessary, top up the oil by way of the filler cap at the rear of the casing. Should too much oil have been added, it should be removed, using a syringe or an empty plastic squeeze pack such as that used for gear oils.

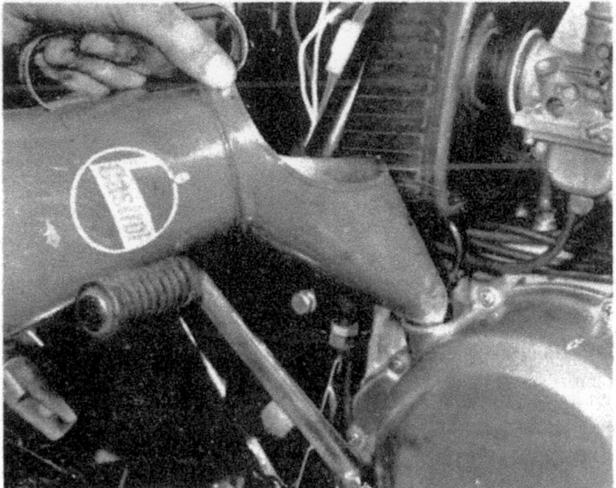

Top up the engine oil level to half way up the window

2 *Tyre pressures*

Check the tyre pressures with a pressure gauge that is known to be accurate. Always check the pressure when the tyres are cold. If the machine has travelled a number of miles, the tyres will have become hot and consequently the pressure will have increased. A false reading wil therefore result.

It is well worth purchasing a small pocket pressure gauge which can be relied on to give consistent readings, and which will remove any reliance on garage forecourt gauges which tend to be less dependable.

Tyre pressures:	*Solo*	*Pillion*
Front:	2·0 kg/cm² (28 psi)	2·0 kg/cm² (28 psi)
Rear:	2·25 kg/cm² (32 psi)	2·50 kg/cm² (36 psi)

Note: pressure readings should be taken when the tyres are cold; i.e. **before** a run.

3 *Hydraulic fluid level*

Check the level of the fluid in the master cylinder reservoir. This may be observed through the translucent side of the reservoir. During normal service, it is unlikely that the hydraulic fluid level will fall dramatically, unless a leak has developed in the system. If this occurs, the fault should be remedied **at once.** The level will fall slowly as the brake linings wear and the fluid deficiency should be corrected, when required. Always use an hydraulic fluid of DOT 3 or SAE J1703 specification, and do not mix different types of fluid, even if the specifications appear the same. This will preclude the possibility of two incompatible fluids being mixed and the resultant chemical reaction damaging the seals.

If the level in the reservoir has been allowed to fall below the specified limit, and air has entered the system, the brake in question must be bled, as described in Chapter 5.

In addition to the above points, a running check of the machine in general should be made. It will be found that conditions such as control cables becoming slack will soon make themselves apparent during riding, necessitating adjustment as soon as possible. The electrical system should also be fully functional, noting that in the UK and in many other countries, it is illegal to use the machine with a defective horn or lights, even if they are not in use.

Weekly or every 250 miles (400 km)

1 *Final drive chain – cleaning and lubrication*

The final drive chain is of the endless type, having no joining link in an effort to eliminate any tendency towards breakage. The rollers are equipped with an 'O' ring at each end which seals the lubricant inside and prevents the ingress of water or abrasive grit. It should not, however, be supposed that the need for lubrication is lessened. On the contrary, frequent but sparse lubrication is essential to minimise wear between the chain and sprockets. This can be accomplished by using one of the aerosol chain lubricants, which have been specifically designed to cling to chains operating at high speed. Note that normal engine or gear oils will be flung off the chain and are not really suitable. The chain and sprockets should be wiped clean before the lubricant is applied, to ensure adequate penetration.

In particularly adverse weather conditions, or when touring, lubrication should be undertaken more frequently.

A final word of caution; The importance of chain lubrication cannot be overstressed in view of the cost of replacement, and the fact that a considerable amount of dismantling work, including swinging arm removal, will need to be undertaken should replacement be necessary.

Adjust the chain after lubrication, so that there is approximately 25 mm (1 in) slack in the middle of the lower run. Always check with the chain at the tightest point as a chain rarely wears evenly during service.

Adjustment is accomplished after placing the machine on the centre stand and slackening the wheel nut, so that the wheel can be drawn backwards by means of the drawbolt adjusters in the fork ends.

The torque arm nuts and the rear brake adjuster must also be slackened during this operation. Adjust the drawbolts an equal amount to preserve wheel alignment. The fork ends are clearly marked with a series of parallel lines above the adjusters, to provide a simple visual check.

2 *Battery – topping up*

A Yuasa YB10L battery, rated at 12 volts 10Ah is fitted as standard and is located in a compartment beneath the dualseat. It is retained by a rubber strap which can be released to allow the battery to be lifted up for examination.

The transparent plastic case of the battery permits the upper and lower levels of the electrolyte to be observed when the battery is lifted from its housing below the dualseat. Maintenance is normally limited to keeping the electrolyte level between the prescribed upper and lower limits and by making sure that the vent pipe is not blocked. The lead plates and their separators can be seen through the translucent case, a further guide to the general condition of the battery.

Unless acid is spilt, as may occur if the machine falls over, the electrolyte should always be topped up with distilled water, to restore the correct level. If acid is spilt on any part of the machine, it should be neutralised with an alkali such as washing soda and washed away with plenty of water, otherwise serious corrosion will occur. Top up with sulphuric acid of the correct specific gravity (1·260 – 1·280) only when spillage has occurred. Check that the vent pipe is well clear of the frame tubes or any of the other cycle parts, for obvious reasons.

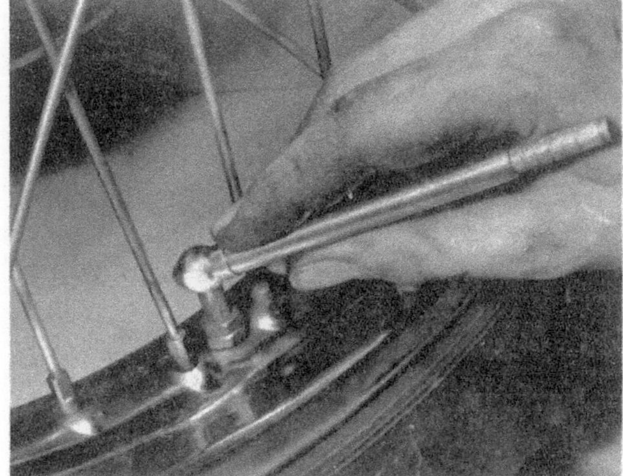

Check the tyre pressures using an accurate gauge

Clean and lubricate the drive chain using aerosol lubricant

Drawbolt assemblies have alignment marks to ensure that the wheel is kept square

3 Control cable lubrication

Apply a few drops of motor oil to the exposed inner portion of each control cable. This will prevent drying-up of the cables between the more thorough lubrication that should be carried out during the 3000 mile/3 monthly service.

4 Safety check

Give the machine a close visual inspection, checking for loose nuts and fittings, frayed control cables etc. Check the tyres for damage, especially splitting of the sidewalls. Remove any stones or other objects caught between the treads. This is particularly important on the front tyre, where rapid deflation due to penetration of the inner tube will almost certainly cause total loss of control.

5 Legal check

Ensure that the lights, horn and flashing indicators function correctly, also the speedometer.

---

3 monthly or every 3000 miles (5000 km)

Carry out the checks listed under the weekly/250 mile heading and then complete the following:

1 Cleaning and adjusting the contact breaker points

Remove the contact breaker inspection cover and gasket. The cover is retained by two screws. Inspect the faces of the two sets of contact breaker points. Deposits due to arcing can be removed while the contact breaker unit is in situ on the machine, using a very fine swiss file or emery paper (No. 400) backed by a thin strip of tin. If the pitting or burning is excessive, the contact breaker unit in question should be removed for points dressing or renewal (see Chapter 3).

Rotate the engine until one set of points is in the fully open position. The correct gap is within the range 0·3 – 0·4 mm (0·012 – 0·016 in). Adjustment is effected by slackening the screw holding the fixed contact breaker point in position and moving the point either closer to or further away with a screwdriver inserted between the small upright post and the slot in the fixed contact plate. Make sure that the points are in the fully open position when this adjustment is made or a false reading will result. When the gap is correct, tighten the screw and recheck.

Repeat the procedure with the other set of points.

2 Checking and resetting the ignition timing

The ignition timing should be checked at the same time as contact breaker adjustment is undertaken. It is desirable to use a multimeter for this purpose, but a battery and bulb arrangement may be used as an alternative.

Referring to the accompanying line drawing, connect 'A' to the terminal post or spring blade of the left-hand contact breaker set (viewed from left-hand side of the machine). 'B' should be connected to a convenient earth point. Using a 17 mm spanner, turn the crankshaft clockwise until the 1·4 F timing mark is visible in the aperture at the top of the backplate. The numbers 1·4 refer to the left-hand contact breaker assembly, also marked 1·4, which triggers the spark on plugs numbers 1 and 4.

If the crankshaft is eased very gradually a few degrees either side of this point, it will be possible to establish the precise point at which the contacts separate. This is indicated by the bulb going out in the case of the battery and bulb apparatus, and by a deflection of the needle where the multimeter is used. Having established the exact point of separation, check that the timing marks still align. If this is not the case, slacken the two screws which retain the left-hand segment, and move this so that separation occurs at the correct point.

Tighten the securing screws, and turn the engine over a few times, then recheck the timing. If all is well, repeat the procedure with the right-hand set of points (2·3), using the appropriate timing mark. Before refitting the contact breaker inspection cover, check that the felt lubricating wick has not dried out. If this should be the case, apply one or two drops of light machine oil, taking care not to over-lubricate it, which would almost certainly cause the contact breaker points to become fouled.

3 Stand pivots and brake fulcrums and pivots

Apply a small amount of EP 90 gear oil to the stand pivot pins, and also to the brake fulcrum and pivots.

4 Control cable lubrication

Lubricate the control cables thoroughly with motor oil or an all-purpose oil. A good method of lubricating the cables is shown in the accompanying illustration, using a plasticine funnel. This method has the disadvantage that the cables usually need removing from the machine. An hydraulic cable oiler which pressurises the lubricant overcomes this problem. Do not lubricate nylon lined cables (which may have been fitted as replacements), as the oil may cause the nylon to swell, thereby causing total cable seizure.

Use feeler gauge to check the contact breaker gaps

Alignment marks are visible through aperture in baseplate

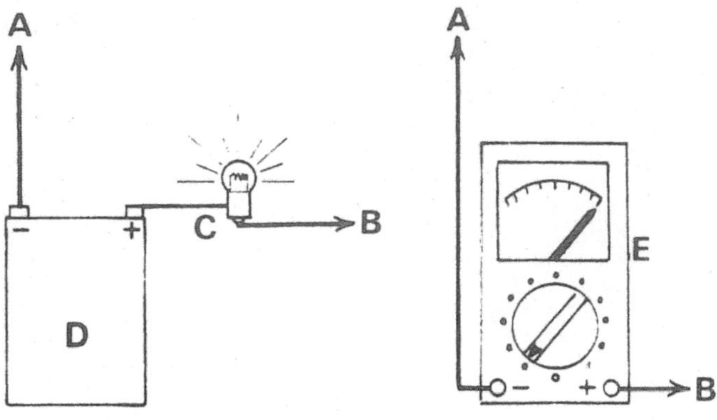

Alternative methods of checking the ignition timing

A: To contact breaker terminal
B: To earth
C: Bulb
D: Battery
E: Multimeter set on resistance

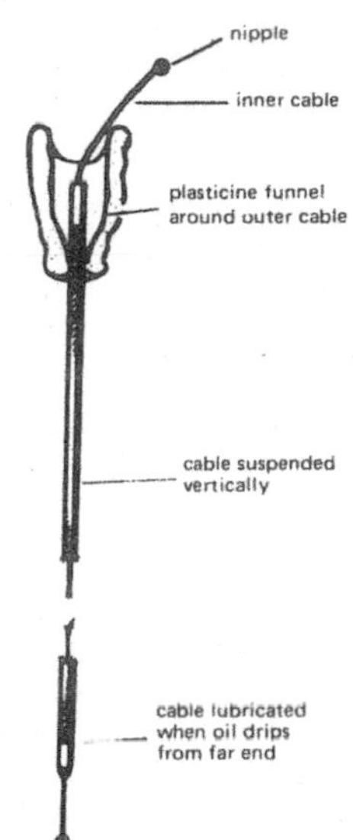

Oiling control cable

5 Changing the engine oil

The oil should be changed with the engine at its normal operating temperature, preferably after a run. This ensures that the oil is relatively thin and will drain more quickly and completely. Obtain a container of at least 3·5 litres (6·2 Imp pints/3·7 US qts) capacity, and arrange it beneath the crankcase drain plug.Slacken and remove the plug, noting that the oil filter cover should be left undisturbed, and allow the oil to drain.

When the crankcase is completely emptied, clean the drain plug orifice and refit the plug, tightening it to 1·3 – 1·7 kg m (9·5 – 12 ft lbs). Remove the filler plug, and add sufficient SAE 10w/40 or 20w/50 motor oil to bring the level half way up the window in the outer cover. This will normally take about 3 litres, the oil filter system requiring about ½ litre if this has been renewed.

6 Camshaft tensioner adjustment

This operation can be performed to some advantage during the checking and adjustment of the ignition timing as much of the work involved is common to both operations. The object is to compensate for wear in the timing chain which would otherwise cause noisy and inaccurate valve operation.

With the contact breaker cover removed, turn the crankshaft forwards so that either the 1·4 or 2·3 timing marks are aligned. This is to arrange the camshafts so that the slackest run is against the tensioner blade. Slacken the locknut and bolt on the side of the tensioner body. This will release the plunger and allow the integral spring to apply the correct amount of pressure to the chain. Finally, tighten the bolt and locknut.

If, after considerable mileages have been covered, this operation fails to restore tension, and chain noise persists, it is indicative of the need to dismantle the tensioner and chain components, and to renew worn parts as necessary.

WARNING: DO NOT USE ELECTRIC STARTER TO ROTATE ENGINE WHEN TENSIONER BOLT IS LOOSE.

7 Clutch adjustment

Accurate adjustment of the clutch is necessary to ensure efficient operation of the whole unit, and to prevent wear of the pushrod assembly. Two adjustment points are available on the cable itself, one at the operating lever, and a second midway along the cable next to the front downtube. The latter should be screwed fully inwards to give the maximum free play in the cable, and the handlebar adjuster set in a central position to give adjustment either way.

Remove the small clutch adjuster cover which is located in the middle of the left-hand outer cover. Slacken the locknut and adjusting screw a few turns, then screw the adjuster inwards until resistance is felt. At this point, all free play in the system is taken up, and the screw must be backed off by a ½ turn. Hold the screw in this position, and tighten the lock nut. Refit the cover, and check the amount of free play in the cable, which should be adjusted at the handlebar lever to give 2 – 3 mm movement.

Set clutch adjustment by way of screw and locknut

8 Valve clearances

It is important that valve clearances be maintained otherwise damage, or at best poor performance and noisy operation, will occur. To gain access to the camshafts, it will be necessary to detach the fuel tank and the H-shaped camshaft cover to expose the two camshafts and their associated components.

Each valve is operated by a bucket-shaped follower which contains a shim to provide the correct clearance between it and the cam lobe. The gap should be measured with the peak of the cam lobe uppermost, at which point it should be possible to insert a feeler gauge between the bucket top and the cam lobe. The specified clearance is between 0·08 mm and 0·18 mm (0·003 – 0·007 in).

To set the camshafts in the correct position, proceed as follows: Turn the crankshaft until the 'Ex' mark on the **exhaust** camshaft is aligned with the cylinder head mating surface, and then check numbers 1 and 3 exhaust valves. Turn the engine so that the 'Ex' mark faces the rear of the cylinder head mating surface, and check valves 2 and 4.

Repeat this procedure on the inlet camshaft, using the 'T' mark aligned first with the rear of the cylinder head surface to check valves 5 and 7, and then at the front to check 6 and 8.

If any of the clearances are outside the specified limits, make a note of the clearance so that the correct size of shim(s) may be obtained. For example; if valve 3 shows a clearance of 0·2 mm, it will be necessary to obtain a shim 0·05 mm thicker which, when fitted, will reduce the clearance to 0·15 mm. The shims range from 2·0 mm to 3·2 mm in 0·05 mm increments.

It will be necessary to remove the camshafts as described in Chapter 1, Section 7, so that the bucket can be withdrawn using a valve grinding stick with a rubber sucker end, or some similar method. The shim may adhere to the inside of the bucket or may stay in position on the valve stem. It should be removed and the new shim placed in position. Reassemble the camshafts and recheck the clearance as described earlier.

In the event that a new shim will not give the required clearance, it is likely that the valve seat and/or valve is in need of renewal. On no account attempt to grind down existing shims or pack them with sheet shim material in an attempt to save the cost of new shims. The risk of failure in service, and the consequent damage to the engine makes this a very false economy. For details of shim sizes refer to the accompanying table of sizes (Fig. RM 2).

9 Carburettors: adjustment

This important operation affects both the performance and fuel consumption of the machine to a marked degree, and should not be ignored despite its complexity. The adjustment operation is described, in greater detail than is possible in this Section, in Chapter 2, Section 6.

10 Air cleaner element: cleaning and renewal

The air cleaner element is located in a compartment underneath the dualseat, and may be withdrawn after the plastic cap has been unscrewed. Wash the element carefully in petrol or a similar solvent, then blow it dry with compressed air. The efficiency of the filter will be slightly impaired with each successive washing, and it should be renewed after every fifth cleaning operation or after every 6000 miles (10 000 km). If there is any sign of damage during cleaning, the element must be renewed as a matter of course.

Six monthly or every 6000 miles (10 000 km)

Carry out the operations listed under the preceding service intervals, then proceed with the following:

1 Brake wear

Check that when applied, the rear brake wear indicator is within the usable range scale marked on the brake plate. The front disc brake pads should also be examined for wear, and to this end are marked with a red line denoting the maximum wear limit. If necessary, change the pads and/or brake shoes, referring to Chapter 5 for details. Look also for signs of staining on the friction material. This may be caused by leakage from the fork leg or from the caliper seals; in either case attention must be given to locating and rectifying the source of the leak.

2 Wheel condition

Check the spoke tension by gently tapping each one with a metal object. A loose spoke is identifiable by the low pitch noise generated. If any spoke needs considerable tightening, it will be necessary to remove the tyre and inner tube in order to file down the protruding spoke end. This will prevent the spoke from chafing through the rim band and piercing the inner tube. Rotate the wheel and test for rim runout. Excessive runout will cause handling problems and should be corrected by tightening or loosening the relevant spokes. Care must be taken, since altering the tension in the wrong spokes may create more problems.

Check clearance between cam lobe and tappet

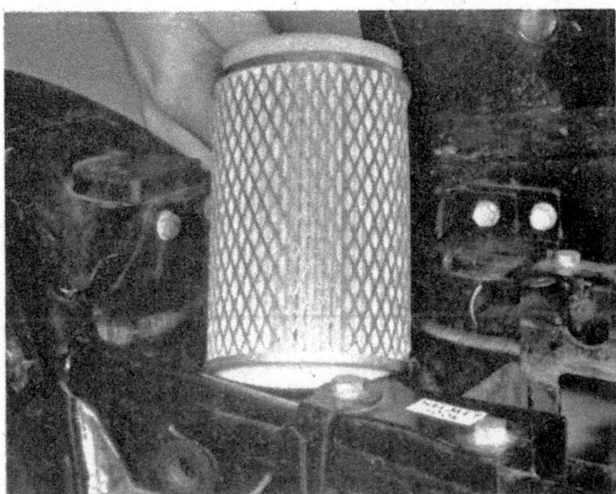

Air filter element is housed in a compartment beneath the seat

VALVE CLEARANCE — PART NUMBER 12037	076	077	078	079	080	081	082	083	084	085	086	087	088	089	090	091	092	093	094	095	096	097	098	099	100
MILIMETERS (PRESENT SHIM SIZE)	2.00	2.05	2.10	2.15	2.20	2.25	2.30	2.35	2.40	2.45	2.50	2.55	2.60	2.65	2.70	2.75	2.80	2.85	2.90	2.95	3.00	3.05	3.10	3.15	3.20
0.00 ~ 0.03				2.00	2.05	2.10	2.15	2.20	2.25	2.30	2.35	2.40	2.45	2.50	2.55	2.60	2.65	2.70	2.75	2.80	2.85	2.90	2.95	3.00	3.05
0.04 ~ 0.08			2.00	2.05	2.10	2.15	2.20	2.25	2.30	2.35	2.40	2.45	2.50	2.55	2.60	2.65	2.70	2.75	2.80	2.85	2.90	2.95	3.00	3.05	3.10
		2.00	2.05	2.10	2.15	2.20	2.25	2.30	2.35	2.40	2.45	2.50	2.55	2.60	2.65	2.70	2.75	2.80	2.85	2.90	2.95	3.00	3.05	3.10	3.15
0.08 ~ 0.18 mm	SPECIFIED CLEARANCE/NO CHANGE REQUIRED																								
0.18 ~ 0.22	2.05	2.10	2.15	2.20	2.25	2.30	2.35	2.40	2.45	2.50	2.55	2.60	2.65	2.70	2.75	2.80	2.85	2.90	2.95	3.00	3.05	3.10	3.15	3.20	
0.23 ~ 0.27	2.10	2.15	2.20	2.25	2.30	2.35	2.40	2.45	2.50	2.55	2.60	2.65	2.70	2.75	2.80	2.85	2.90	2.95	3.00	3.05	3.10	3.15	3.20		
0.28 ~ 0.32	2.15	2.20	2.25	2.30	2.35	2.40	2.45	2.50	2.55	2.60	2.65	2.70	2.75	2.80	2.85	2.90	2.95	3.00	3.05	3.10	3.15	3.20			
0.33 ~ 0.37	2.20	2.25	2.30	2.35	2.40	2.45	2.50	2.55	2.60	2.65	2.70	2.75	2.80	2.85	2.90	2.95	3.00	3.05	3.10	3.15	3.20				
0.38 ~ 0.42	2.25	2.30	2.35	2.40	2.45	2.50	2.55	2.60	2.65	2.70	2.75	2.80	2.85	2.90	2.95	3.00	3.05	3.10	3.15	3.20					
0.43 ~ 0.47	2.30	2.35	2.40	2.45	2.50	2.55	2.60	2.65	2.70	2.75	2.80	2.85	2.90	2.95	3.00	3.05	3.10	3.15	3.20						
0.48 ~ 0.52	2.35	2.40	2.45	2.50	2.55	2.60	2.65	2.70	2.75	2.80	2.85	2.90	2.95	3.00	3.05	3.10	3.15	3.20							
0.53 ~ 0.57	2.40	2.45	2.50	2.55	2.60	2.65	2.70	2.75	2.80	2.85	2.90	2.95	3.00	3.05	3.10	3.15	3.20								
0.58 ~ 0.62	2.45	2.50	2.55	2.60	2.65	2.70	2.75	2.80	2.85	2.90	2.95	3.00	3.05	3.10	3.15	3.20									
0.63 ~ 0.67	2.50	2.55	2.60	2.65	2.70	2.75	2.80	2.85	2.90	2.95	3.00	3.05	3.10	3.15	3.20										
0.68 ~ 0.72	2.55	2.60	2.65	2.70	2.75	2.80	2.85	2.90	2.95	3.00	3.05	3.10	3.15	3.20											
0.73 ~ 0.77	2.60	2.65	2.70	2.75	2.80	2.85	2.90	2.95	3.00	3.05	3.10	3.15	3.20												
0.78 ~ 0.82	2.65	2.70	2.75	2.80	2.85	2.90	2.95	3.00	3.05	3.10	3.15	3.20													
0.83 ~ 0.87	2.70	2.75	2.80	2.85	2.90	2.95	3.00	3.05	3.10	3.15	3.20														
0.88 ~ 0.92	2.75	2.80	2.85	2.90	2.95	3.00	3.05	3.10	3.15	3.20															
0.93 ~ 0.97	2.80	2.85	2.90	2.95	3.00	3.05	3.10	3.15	3.20																
0.98 ~ 1.02	2.85	2.90	2.95	3.00	3.05	3.10	3.15	3.20																	
1.03 ~ 1.07	2.90	2.95	3.00	3.05	3.10	3.15	3.20																		
1.08 ~ 1.12	2.95	3.00	3.05	3.10	3.15	3.20																			
1.13 ~ 1.17	3.00	3.05	3.10	3.15	3.20																				
1.18 ~ 1.22	3.05	3.10	3.15	3.20																					
1.23 ~ 1.27	3.10	3.15	3.20																						
1.28 ~ 1.32	3.15	3.20																							
1.33 ~ 1.37	3.20																								

INSTALL THIS SHIM

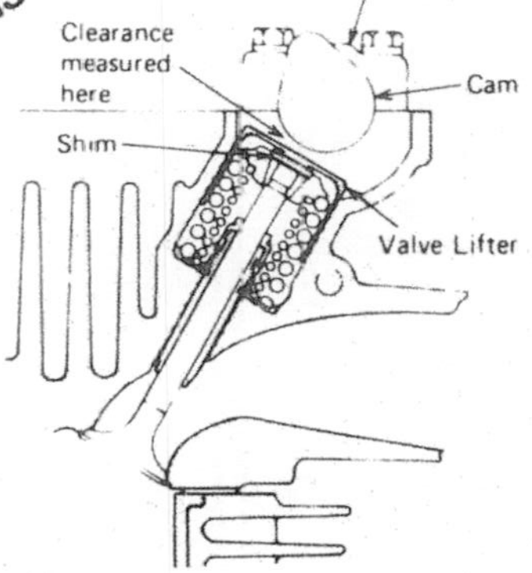

Shim selection table

3 *Changing the oil filter element*

Change the engine oil as described under the 3 monthly 3000 mile heading. In addition, remove and renew the oil filter element which is housed in a chamber in the underside of the crankcase.

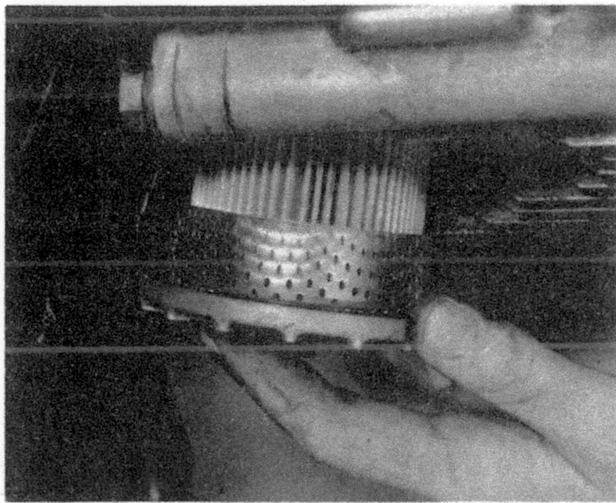

Oil filter assembly screws into the bottom of the crankcase

4 *Changing the front fork damping oil*

Place the machine on the centre stand so that the front wheel is clear of the ground. Place wooden blocks below the crankcase in order to prevent the machine from tipping forward. Loosen and remove the chrome cap bolts. Unscrew the drain plug from each fork leg, located directly above the wheel spindle, and allow the damping fluid to drain into a suitable container. This is accomplished most easily if the legs are attended to in turn. Take care not to spill any fluid onto the brake disc or tyre. The forks may be pumped up and down slowly to expel any remaining fluid. Refit and tighten the drain plugs. Refill each fork leg with 183 – 191 cc of SAE 15W engine oil, or a good quality fork oil. If a straight grade fork oil is chosen, either SAE 10, 20 or 30 may be used, depending on choice. The thicker the oil the heavier the damping will be. Refit and tighten the chrome cap bolts.

5 *General checks*

In addition to the above operations, the various frame and engine fittings should be checked for tightness and lubricated where necessary. It is recommended that the hydraulic fluid be changed, by following the brake bleeding procedure in Chapter 5. Check for free play in the steering head bearings, and adjust if necessary, following the sequence in Chapter 4.

Every two years or 12 000 miles (20 000 km)

Remove the front wheel and repack the front wheel bearings with high melting point grease. At the same time, remove and re-grease the speedometer drive assembly. The rear wheel bearings and brake cam should also be re-greased. Details of dismantling and reassembly will be found in Chapter 5. The steering head should also be dismantled for examination and re-lubrication as described in Chapter 4.

Quick glance maintenance adjustments and capacities

Engine/gearbox unit	3·0 litres (5·3 imp pints/3·1 US quarts) SAE10/W40, 10/W50 or 20/W50 engine oil (If filter is renewed, an extra ½ litre (0·9 imp pints/ 0·5 US quarts) will be required.	
Front forks	183 – 191 cc per leg SAE 15W engine oil or Fork oil	
Contact breaker gap	0·3 – 0·4 mm (0·012 – 0·016 in)	
Sparking plug gap	0·7 – 0·8 mm (0·027 – 0·031 in)	
Tyre pressures:	**Front**	**Rear**
Solo	2·0 kg/cm² (28 psi)	2·25 kg/cm² (32 psi)
Pillion, or high speed	2·0 kg/cm² (28 psi)	2·50 kg/cm² (36 psi)

Recommended lubricants

Engine/gearbox unit	SAE 10/W40, 10/W50 or 20/W50 engine oil
Front forks	SAE 10/W20 engine oil, or Fork oil
Final drive chain	Aerosol chain lubricant
Pivot points	High melting point grease or gear oil
Wheel bearings	High melting point grease

Safety first!

Professional motor mechanics are trained in safe working procedures. However enthusiastic you may be about getting on with the job in hand, do take the time to ensure that your safety is not put at risk. A moment's lack of attention can result in an accident, as can failure to observe certain elementary precautions.

There will always be new ways of having accidents, and the following points do not pretend to be a comprehensive list of all dangers; they are intended rather to make you aware of the risks and to encourage a safety-conscious approach to all work you carry out on your vehicle.

Essential DOs and DON'Ts

DON'T start the engine without first ascertaining that the transmission is in neutral.

DON'T suddenly remove the filler cap from a hot cooling system – cover it with a cloth and release the pressure gradually first, or you may get scalded by escaping coolant.

DON'T attempt to drain oil until you are sure it has cooled sufficiently to avoid scalding you.

DON'T grasp any part of the engine, exhaust or silencer without first ascertaining that it is sufficiently cool to avoid burning you.

DON'T allow brake fluid or antifreeze to contact the machine's paintwork or plastic components.

DON'T syphon toxic liquids such as fuel, brake fluid or antifreeze by mouth, or allow them to remain on your skin.

DON'T inhale dust – it may be injurious to health (see *Asbestos* heading).

DON'T allow any spilt oil or grease to remain on the floor – wipe it up straight away, before someone slips on it.

DON'T use ill-fitting spanners or other tools which may slip and cause injury.

DON'T attempt to lift a heavy component which may be beyond your capability – get assistance.

DON'T rush to finish a job, or take unverified short cuts.

DON'T allow children or animals in or around an unattended vehicle.

DON'T inflate a tyre to a pressure above the recommended maximum. Apart from overstressing the carcase and wheel rim, in extreme cases the tyre may blow off forcibly.

DO ensure that the machine is supported securely at all times. This is especially important when the machine is blocked up to aid wheel or fork removal.

DO take care when attempting to slacken a stubborn nut or bolt. It is generally better to pull on a spanner, rather than push, so that if slippage occurs you fall away from the machine rather than on to it.

DO wear eye protection when using power tools such as drill, sander, bench grinder etc.

DO use a barrier cream on your hands prior to undertaking dirty jobs – it will protect your skin from infection as well as making the dirt easier to remove afterwards; but make sure your hands aren't left slippery. Note that long-term contact with used engine oil can be a health hazard.

DO keep loose clothing (cuffs, tie etc) and long hair well out of the way of moving mechanical parts.

DO remove rings, wristwatch etc, before working on the vehicle – especially the electrical system.

DO keep your work area tidy – it is only too easy to fall over articles left lying around.

DO exercise caution when compressing springs for removal or installation. Ensure that the tension is applied and released in a controlled manner, using suitable tools which preclude the possibility of the spring escaping violently.

DO ensure that any lifting tackle used has a safe working load rating adequate for the job.

DO get someone to check periodically that all is well, when working alone on the vehicle.

DO carry out work in a logical sequence and check that everything is correctly assembled and tightened afterwards.

DO remember that your vehicle's safety affects that of yourself and others. If in doubt on any point, get specialist advice.

IF, in spite of following these precautions, you are unfortunate enough to injure yourself, seek medical attention as soon as possible.

Asbestos

Certain friction, insulating, sealing, and other products – such as brake linings, clutch linings, gaskets, etc – contain asbestos. *Extreme care must be taken to avoid inhalation of dust from such products since it is hazardous to health.* If in doubt, assume that they *do* contain asbestos.

Fire

Remember at all times that petrol (gasoline) is highly flammable. Never smoke, or have any kind of naked flame around, when working on the vehicle. But the risk does not end there – a spark caused by an electrical short-circuit, by two metal surfaces contacting each other, by careless use of tools, or even by static electricity built up in your body under certain conditions, can ignite petrol vapour, which in a confined space is highly explosive.

Always disconnect the battery earth (ground) terminal before working on any part of the fuel or electrical system, and never risk spilling fuel on to a hot engine or exhaust.

It is recommended that a fire extinguisher of a type suitable for fuel and electrical fires is kept handy in the garage or workplace at all times. Never try to extinguish a fuel or electrical fire with water.

Note: *Any reference to a 'torch' appearing in this manual should always be taken to mean a hand-held battery-operated electric lamp or flashlight. It does* **not** *mean a welding/gas torch or blowlamp.*

Fumes

Certain fumes are highly toxic and can quickly cause unconsciousness and even death if inhaled to any extent. Petrol (gasoline) vapour comes into this category, as do the vapours from certain solvents such as trichloroethylene. Any draining or pouring of such volatile fluids should be done in a well ventilated area.

When using cleaning fluids and solvents, read the instructions carefully. Never use materials from unmarked containers – they may give off poisonous vapours.

Never run the engine of a motor vehicle in an enclosed space such as a garage. Exhaust fumes contain carbon monoxide which is extremely poisonous; if you need to run the engine, always do so in the open air or at least have the rear of the vehicle outside the workplace.

The battery

Never cause a spark, or allow a naked light, near the vehicle's battery. It will normally be giving off a certain amount of hydrogen gas, which is highly explosive.

Always disconnect the battery earth (ground) terminal before working on the fuel or electrical systems.

If possible, loosen the filler plugs or cover when charging the battery from an external source. Do not charge at an excessive rate or the battery may burst.

Take care when topping up and when carrying the battery. The acid electrolyte, even when diluted, is very corrosive and should not be allowed to contact the eyes or skin.

If you ever need to prepare electrolyte yourself, always add the acid slowly to the water, and never the other way round. Protect against splashes by wearing rubber gloves and goggles.

Mains electricity and electrical equipment

When using an electric power tool, inspection light etc, always ensure that the appliance is correctly connected to its plug and that, where necessary, it is properly earthed (grounded). Do not use such appliances in damp conditions and, again, beware of creating a spark or applying excessive heat in the vicinity of fuel or fuel vapour. Also ensure that the appliances meet the relevant national safety standards.

Ignition HT voltage

A severe electric shock can result from touching certain parts of the ignition system, such as the HT leads, when the engine is running or being cranked, particularly if components are damp or the insulation is defective. Where an electronic ignition system is fitted, the HT voltage is much higher and could prove fatal.

Working conditions and tools

When a major overhaul is contemplated, it is important that a clean, well-lit working space is available, equipped with a workbench and vice, and with space for laying out or storing the dismantled assemblies in an orderly manner where they are unlikely to be disturbed. The use of a good workshop will give the satisfaction of work done in comfort and without haste, where there is little chance of the machine being dismantled and reassembled in anything other than clean surroundings. Unfortunately, these ideal working conditions are not always practicable and under these latter circumstances when improvisation is called for, extra care and time will be needed.

The other essential requirement is a comprehensive set of good quality tools. Quality is of prime importance since cheap tools will prove expensive in the long run if they slip or break when in use, causing personal injury or expensive damage to the component being worked on. A good quality tool will last a long time, and more than justify the cost.

For practically all tools, a tool factor is the best source since he will have a very comprehensive range compared with the average garage or accessory shop. Having said that, accessory shops often offer excellent quality tools at discount prices, so it pays to shop around. There are plenty of tools around at reasonable prices, but always aim to purchase items which meet the relevant national safety standards. If in doubt, seek the advice of the shop proprietor or manager before making a purchase.

The basis of any tool kit is a set of open-ended spanners, which can be used on almost any part of the machine to which there is reasonable access. A set of ring spanners makes a useful addition, since they can be used on nuts that are very tight or where access is restricted. Where the cost has to be kept within reasonable bounds, a compromise can be effected with a set of combination spanners – open-ended at one end and having a ring of the same size on the other end. Socket spanners may also be considered a good investment, a basic 3/8 in or 1/2 in drive kit comprising a ratchet handle and a small number of socket heads, if money is limited. Additional sockets can be purchased, as and when they are required. Provided they are slim in profile, sockets will reach nuts or bolts that are deeply recessed. When purchasing spanners of any kind, make sure the correct size standard is purchased. Almost all machines manufactured outside the UK and the USA have metric nuts and bolts, whilst those produced in Britain have BSF or BSW sizes. The standard used in USA is AF, which is also found on some of the later British machines. Others tools that should be included in the kit are a range of crosshead screwdrivers, a pair of pliers and a hammer.

When considering the purchase of tools, it should be remembered that by carrying out the work oneself, a large proportion of the normal repair cost, made up by labour charges, will be saved. The economy made on even a minor overhaul will go a long way towards the improvement of a toolkit.

In addition to the basic tool kit, certain additional tools can prove invaluable when they are close to hand, to help speed up a multitude of repetitive jobs. For example, an impact screwdriver will ease the removal of screws that have been tightened by a similar tool, during assembly, without a risk of damaging the screw heads. And, of course, it can be used again to retighten the screws, to ensure an oil or airtight seal results. Circlip pliers have their uses too, since gear pinions, shafts and similar components are frequently retained by circlips that are not too easily displaced by a screwdriver. There are two types of circlip pliers, one for internal and one for external circlips. They may also have straight or right-angled jaws.

One of the most useful of all tools is the torque wrench, a form of spanner that can be adjusted to slip when a measured amount of force is applied to any bolt or nut. Torque wrench settings are given in almost every modern workshop or service manual, where the extent to which a complex component, such as a cylinder head, can be tightened without fear of distortion or leakage. The tightening of bearing caps is yet another example. Overtightening will stretch or even break bolts, necessitating extra work to extract the broken portions.

As may be expected, the more sophisticated the machine, the greater is the number of tools likely to be required if it is to be kept in first class condition by the home mechanic. Unfortunately there are certain jobs which cannot be accomplished successfully without the correct equipment and although there is invariably a specialist who will undertake the work for a fee, the home mechanic will have to dig more deeply in his pocket for the purchase of similar equipment if he does not wish to employ the services of others. Here a word of caution is necessary, since some of these jobs are best left to the expert. Although an electrical multimeter of the AVO type will prove helpful in tracing electrical faults, in inexperienced hands it may irrevocably damage some of the electrical components if a test current is passed through them in the wrong direction. This can apply to the synchronisation of twin or multiple carburettors too, where a certain amount of expertise is needed when setting them up with vacuum gauges. These are, however, exceptions. Some instruments, such as a strobe lamp, are virtually essential when checking the timing of a machine powered by CDI ignition system. In short, do not purchase any of these special items unless you have the experience to use them correctly.

Although this manual shows how components can be removed and replaced without the use of special service tools (unless absolutely essential), it is worthwhile giving consideration to the purchase of the more commonly used tools if the machine is regarded as a long term purchase Whilst the alternative methods suggested will remove and replace parts without risk of damage, the use of the special tools recommended and sold by the manufacturer will invariably save time.

Chapter 1 Engine, clutch and gearbox

Contents

Specifications

Engine

Type	4 cylinder transverse dohc in-line air cooled four stroke
Bore	62 mm
Stroke	54 mm
Displacement	652 cc
Compression ratio	9.5 : 1
Maximum horsepower	64 hp @ 8500 rpm
Maximum torque	5.8 kgm @ 7000 rpm
Crankcase	Aluminium alloy, horizontally split incorporating gearbox and oil sump
Cylinder block	Aluminium alloy, steel liners
Cylinder head	Aluminium alloy

Pistons

Type	Aluminium alloy, solid skirt
Piston ring groove width	Top 1.23 – 1.25 mm (wear limit 1.33 mm) 2nd 1.22 – 1.24 mm (wear limit 1.32 mm) Oil 2.51 – 2.53 mm (wear limit 2.60 mm)
Oversizes available	0.5 and 1.0 mm

Piston rings

Number per piston	Two compression, one oil control
End gap (free)	8 mm (wear limit 5.0 mm)
End gap (installed)	0.15 – 0.30 mm (wear limit 0.70 mm)
Ring thickness	Top 1.17 – 1.19 mm (wear limit 1.10 mm) 2nd 1.17 – 1.19 mm (wear limit 1.10 mm) Oil N/A
Piston ring to groove clearance	Top 0.04 – 0.08 mm (wear limit 0.15 mm) 2nd 0.03 – 0.07 mm (wear limit 0.15 mm) Oil N/A

Cylinder bores

Standard bore diameter	61.995 – 62.003 mm
Wear limit	62.100 mm
Piston/bore clearance	0.032 – 0.055 mm (0.0013 – 0.0022 in)

Small end assembly

Gudgeon pin OD	14.994 – 14.998 mm (wear limit 14.96 mm)
Piston boss diameter	15.004 – 15.009 mm (wear limit 15.07 mm)
Small end bush diameter	15.003 – 15.014 mm (wear limit 15.05 mm)
Piston boss to gudgeon pin clearance	0.006 – 0.015 mm (0.0002 – 0.0006 in)
Gudgeon pin to small end bush clearance	0.005 – 0.020 mm (0.0002 – 0.0008 in)

Big end assembly

Bearing to journal clearance	0.041 – 0.067 mm (wear limit 0.1 mm) (0.0016 – 0.0026 in) (0.0039 in)
Big end side clearance	0.15 – 0.25 mm (wear limit 0.45 mm) (0.0059 – 0.0098 in) (0.0177 in)
Big end journal diameter	
Unmarked	34.984 – 34.994 mm (1.3773 – 1.3777 in)
Marked '1'	34.995 – 35.000 mm (1.3778 – 1.3780 in)
Wear limit	34.97 mm (1.3768 in)
Connecting rod big end diameter	
Marked '1'	38.009 – 38.016 mm (1.4964 – 1.4967 in)
Marked '2'	38.000 – 38.008 mm (1.4961 – 1.4964 in)
Main bearing journal diameter	35.984 – 36.000 mm (1.4167 – 1.4173 in)
Wear limit	35.94 mm (1.4150 in)
Main bearing/journal clearance	0.034 – 0.076 mm (0.0013 – 0.0030 in)
Wear limit	0.11 mm (0.0043 in)

Valves

Valve stem diameter, inlet	6.965 – 6.980 mm (0.2742 – 0.2748 in)
exhaust	6.950 – 6.970 mm (0.2736 – 0.2744 in)
Wear limit, inlet	6.90 mm (0.2716 in)
exhaust	6.89 mm (0.2712 in)
Valve clearance (cold engine)	0.08 – 0.18 mm (0.003 – 0.007 in)

Valve timing

Inlet opens at	22° BTDC
Inlet closes at	52° ABDC
Duration	254°
Exhaust opens at	60° BBDC
Exhaust closes at	52° ATDC
Duration	260°

Clutch

Number of plain plates	6
Number of friction plates	7
Friction plate thickness	3.7 – 3.9 mm (0.146 – 0.154 in)
Wear limit	3.5 mm (0.138 in)
Friction plate warpage	less than 0.15 mm (0.0059 in) nominal less than 0.30 mm (0.0118 in) wear limit
Plain plate warpage	less than 0.12 mm (0.0047 in) nominal less than 0.40 mm (0.0157 in) wear limit

Gearbox

Type	5 speed constant mesh
Gear ratio	1st 2.33 : 1 (35/15)
	2nd 1.63 : 1 (31/19)
	3rd 1.27 : 1 (28/22)
	4th 1.04 : 1 (26/25)
	Top 0.89 : 1 (24/27)
Primary reduction ratio	2.55 (27/23 x 63/29)
Final reduction ratio	2.63 (42/16)
Overall drive ratio	5.95 (in top gear)

Torque wrench settings

Component	Nut/ bolt size	Quantity	kg m	ft lbs	in lbs	locking agent required
Cylinder head nuts	10 mm	12	3.5 – 4.0	25.0 – 29.0	–	No
Cylinder head bolts	6 mm	2	2.2 – 2.8	16.0 – 20.0	–	No
Cylinder head cover bolts	6 mm	24	0.7 – 0.9	–	61 – 78	No
Camshaft cap bolts	6 mm	16	1.1 – 1.3	–	95 – 113	No
Cam chain tensioner adjustment bolt	6 mm	1	0.9 – 1.1	–	78 – 95	No
Cam chain tensioner body bolts	6 mm	2	0.7 – 0.9	–	61 – 78	Yes
Cam chain jockey wheel spindle bolt	6 mm	1	0.8 – 1.0	–	69 – 87	Yes
Camshaft sprocket mounting bolts	6 mm	4	1.3 – 1.7	9.5 – 12.0	–	Yes
Crankcase bolts:						
upper	6 mm	12	0.9 – 1.1	–	78 – 95	Yes
lower	6 mm	8	0.9 – 1.1	–	78 – 95	Yes
lower	8 mm	10	2.5 – 3.0	18 – 22	–	Yes
Crankcase drain plug	12 mm	1	1.3 – 1.7	9.5 – 12.0	–	No
Sump mounting bolts	6 mm	15	0.7 – 0.9	–	61 – 78	No
Engine mounting bolts	10 mm	4	3.4 – 4.6	25 – 33	–	No
	8 mm	6	2.0 – 2.8	14.5 – 20.0	–	No
Gearbox sprocket guard bolts	6 mm	3	N/A	N/A	N/A	Yes
Gearbox sprocket nut	20 mm	1	7.5 – 8.5	54 – 61	–	No
Clutch centre nut	20 mm	1	12 – 15	87 – 100	–	No
Clutch spring bolts	6 mm	5	0.9 – 1.1	–	78 – 95	No
Oil pressure switch	–	1	1.3 – 1.7	9.5 – 12.0	–	No
Oil pressure relief valve	12 mm	1	1.3 – 1.7	9.5 – 12.0	–	Yes
Oil pump mounting bolts	6 mm	2	–	–	–	Yes
Screw	–	1	–	–	–	Yes
ATU bolt	8 mm	1	2.3 – 2.7	16.5 – 19.5	–	No

1 General description

The engine unit fitted to the Kawasaki Z650 series is of the 4 cylinder in-line type, fitted transversely across the frame. The valves are operated by double overhead camshafts driven off the crankshaft by a centre chain. The two camshafts are located in the cylinder head casting, and the camshaft chain drive operates through a cast-in tunnel between the four cylinders. Adjustment of the chain is effected by a chain tensioner, fitted to the rear of the cylinder block.

The engine/gear unit is of aluminium alloy construction, with the crankcase divided horizontally.

The Z650 series has a wet sump, pressure fed lubrication system, which incorporates a gear driven oil pump, an oil filter, a safety by-pass valve, and an oil pressure switch.

Oil vapours created in the crankcase are vented through an oil breather to the air cleaner hose where they are recirculated into the crankcase, providing an oil tight system.

The oil pump is a twin shaft dual rotor unit, which is driven off the crankshaft by a gear.

An oil strainer is fitted to the intake side of the oil pump, which serves to protect the pump mechanism from any impurities in the oil that might cause damage.

The oil filter unit, which is housed in the sump, is an alloy canister with a paper element. As the oil filter becomes clogged with impurities, its ability to operate efficiently is reduced, and when it becomes so clogged that it begins to impede the oil flow, the by-pass valve opens and routes the oil around the filter. This of course results in unfiltered oil being circulated throughout the engine, a condition that will be avoided if the filter element is changed at the prescribed intervals.

The lubrication flow is as follows: Oil is drawn from the sump through the oil strainer to the pump, then it passes through the oil filter (or around it if the by-pass valve is in operation) to the pipe in which the oil pressure switch is mounted. It is then routed through three branch systems. The first system lubricates the crankshaft main bearings and crankpins. The oil is thrown by the crankshaft's rotating motion onto the cylinder walls providing the splash lubrication for the pistons. The oil then drips down into the sump, to be recirculated.

The second system lubricates the cylinder head assembly. Oil flows up through passages in the cylinder block, through the camshaft bushes, down over the cams, through the cam lifters (or tappets) and back to the sump by way of holes in the base of the tappets, and the cam chain tunnel in the cylinder head.

The third system feeds the transmission bearings and then drains back to the sump for recirculation.

The engine is built in unit with the gearbox. This means that when the engine is completely dismantled, the clutch and gearbox are dismantled too. This task is made easy by arranging the crankcase to separate horizontally.

2 Operations with the engine/gearbox unit in the frame

1 It is not necessary to remove the engine from the frame to carry out certain operations; in fact it can be an advantage. Tasks that can be carried out with the engine in place are as follows:

a) Removal and replacement of the clutch
b) Removal and replacement of the flywheel generator.
c) Removal and replacement of the generator rotor.
d) Removal and replacement of the carburettors.
e) Removal and replacement of the starter motor.
f) Removal and replacement of the secondary shaft components.

2 When several tasks have to be undertaken simultaneously, it will probably be advantageous to remove the complete engine unit from the frame. This gives the advantage of much better access and more working space.

3 Operations with the engine/gearbox unit removed from the frame

a) Removal and replacement of the cylinder head unit.
b) Removal and replacement of the cylinder block.
c) Removal and replacement of the pistons.
d) Removal and replacement of the crankshaft assembly.
e) Removal and replacement of the main bearings.
f) Removal and replacement of the gear clusters and selectors.
g) Removal and replacement of the kickstart mechanism, gearbox bearings, and gear change mechanism.

4 Method of engine/gearbox removal

As mentioned previously, the engine and gearbox are built in unit and it is necessary to remove the complete unit to gain access to either assembly.

The engine unit is secured to the frame with 10 mounting bolts. After these have been removed, and the necessary electrical connections disconnected, together with the carburettor fuel pipes, plug leads and exhaust system, the engine is ready for removal. Dismantling of the engine unit can only be accomplished after the engine unit has been removed from the frame and refitting cannot take place until the engine unit has been reassembled.

5 Removing the engine/gearbox unit

1 Place the machine firmly on its centre stand so that it stands on a smooth, level surface. The ideal position for working is to place the machine on a stout wooden stand about 18 inches high, resting on its centre stand.
2 Make sure you have a clean, well-lit place to work in and a good set of tools. You will need at least three sizes of crosshead (Phillips) screwdrivers, small, medium and large, and plenty of clean lint free rag.
3 Remove the oil sump plug, oil filter cover and element, and drain the oil into a suitable tray. Approximately 3.5 litres of oil will drain off.
4 Disconnect and remove the battery. The battery is located beneath the dualseat, in a cradle compartment. It should be lifted straight up, being careful not to spill the contents.
5 Switch off the petrol tap and disconnect the petrol feed pipe from the stub on the tap body. The tank is held by a rubber clip at the rear that engages in a lip welded onto the rear of the tank. Unhook the rubber band and pull the fuel tank off, toward the rear.
6 Remove the finned clamps from all four exhaust pipes. They are held by two nuts per clamp, secured to studs fitted into the cylinder head. It is a good idea to soak these nuts in penetrating oil before undoing them, to safeguard against breakage of the studs in the cylinder head. Slacken the clamps which hold the exhaust balance pipe, beneath the crankcase, to the system at each side of the machine. The two silencer mounting bolts can now be released, and the system pulled clear of the machine. The balance pipe will remain on one half of the system.
7 Remove the two side panels to expose the electrical components housed either side of the frame. Disengage the rubber-mounted starter solenoid from the left-hand side of the machine, which will then allow the air cleaner trunking mounting bolt to be released. The bracket to which this attaches will disengage from the plastic casing, and should be removed for safe keeping. Detach the right-hand mounting in a similar manner.
8 The carburettors can now be removed as a complete unit. It should be noted that this is not an easy task, and it is preferable to have an assistant to hand to help manoeuvre the assembly clear. Slacken the clamps which retain each carburettor to its rubber inlet stub, and roll the spring bands up the hoses which connect the instruments to the air cleaner trunking. Pull the carburettor assembly back away from the inlet stubs and twist it to disengage the instruments' mouths. This operation calls for a fair degree of patience if damage is to be avoided. Once free, lift the assembly away, complete with the various drain and breather hoses. Disconnect the throttle cables, which will be more accessible as the assembly is withdrawn.
9 Remove the two bolts which secure the starter motor cover, and lift the cover away. Remove the left-hand footrest, gearchange pedal and the left-hand casing securing bolts, then remove the casing. Knock back the tab washer which locks the gearbox sprocket nut. Apply the rear brake to immobilise the sprocket, then slacken the securing nut. The sprocket may now be slid off the shaft and disengaged from the chain, the latter remaining in position on the machine.
10 Disconnect the neutral switch lead and disconnect the clutch cable to release the outer casing. Slacken and remove the starter motor mounting bolts, then pull the motor clear of the unit and detach the lead from the solenoid. Moving to the right-hand side of the machine, disconnect the rear brake light switch leads, and the earth cable to the rear of the crankcase. Unhook the spring which operates the brake switch. Disconnect the contact breaker leads (coloured black, green and blue/red). Remove the right-hand footrest. Slacken the rear brake adjusting nut, then detach the rear brake pedal.
11 Unscrew the gland nut which retains the tachometer drive cable to the cylinder head. Disengage the cable and position it out of the way. Remove the sparking plug caps, noting that the leads are numbered to avoid confusion during reassembly. Check around the unit carefully to ensure that no cables or leads have been left connected, and that no other component or fitting is likely to impede engine removal. Remove the ignition coils and lodge them away from the unit on top of the frame top tubes.
12 Slacken and remove the nuts from the engine mounting bolts in the sequence shown in the accompanying photograph. Carefully remove the mounting bolts and engine plates, supporting the unit when necessary to aid their removal. The unit will now sit on the lower part of the cradle. The engine/gearbox unit is heavy, bulky and awkward, and will require two persons at least if damage to the machine and its owner is to be avoided. Lift the unit up squarely and move it to the right, checking that the crankcase clears the front and rear lower mounting bolt lugs. Lift the right side of the unit so that the sump will clear the cradle, then ease the unit out to the right. Once clear, the unit can be manhandled onto a suitably sturdy bench, leaving the now exhausted removal party to take a well earned rest.

5.8 Remove carburettors as a unit, detaching opening and closing cables

5.9a Remove the gearchange pedal and left-hand footrest

5.9b Lift away the outer cover, and disconnect clutch cable

5.9c Lock rear wheel with brake, then remove sprocket nut

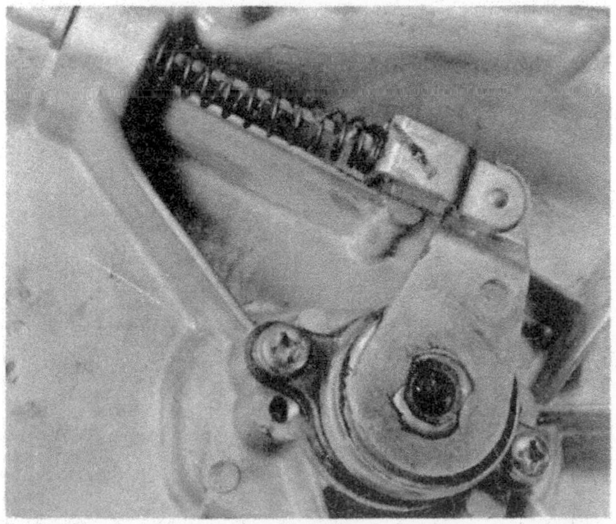
5.10a Clutch cable may be disconnected at lever and removed with cover

5.10b Separate various leads and connector blocks

5.11 Tachometer drive cable is retained by knurled ring

5.12 Engine mounting bolts should be removed and tightened in the sequence shown above

6 Dismantling the engine/gearbox: general

1 Before commencing work on the engine unit, the external surfaces should be cleaned thoroughly. A motorcycle engine has very little protection from the hazards of road grit and other foreign matter, due to it having to be constructed to take advantage of air cooling.

2 There are a number of proprietary cleaning solvents on the market including Jizer and Gunk. It is best to soak the parts in one of these solvents, using a cheap paint brush. Allow the solvent to penetrate the dirt, and afterwards wash down with water, making sure not to let water penetrate the electrical system or get into the engine, as many parts are now more exposed.

3 Have a good set of tools ready, including a set of open ring metric spanners (these have a ring one end and are open ended at the other end), a few metric socket spanners of the smaller sizes, and an impact screwdriver with a selection of bits for the crosshead screws. If one is not available, a crosshead screwdriver with a T handle fitted can sometimes be used as a substitute. Work on a clean surface and have a supply of clean lint free rag available.

4 Never use force to remove any stubborn part unless specific mention is made of this requirement in the text. There is invariably good reason why a part is difficult to remove, often because the dismantling procedure has been tackled out of sequence.

7 Dismantling the engine/gearbox: removing the camshafts, cylinder head and cylinder block

1 Remove the H-shaped camshaft cover and place it to one side. Slacken the camshaft chain tensioner mounting bolts, and remove the unit, allowing the chain to become slack. Remove the screw which retains the tachometer drive pinion, and withdraw the latter from the cylinder head. Slacken and remove the socket screws which retain the chain guide sprocket to the centre of the cylinder head, and lift the assembly away, together with the metal/rubber mounting blocks.

2 Slacken the camshaft cap bolts, and lift off the caps. The camshafts may now be lifted in turn, and threaded out of position. Take care not to let the chain drop down into the crankcase, unless a complete stripdown is intended. There is no need to mark the camshafts, as the exhaust camshaft incorporates the tachometer driving worm and is thus easily identified.

3 The cylinder head is retained by twelve 10 mm nuts, and two 6 mm bolts which are recessed into the cam chain tunnel. Slacken and remove these in the reverse order of the tightening sequence (See Fig. 1.10) to avoid the risk of warping the head casting. The cylinder head should now lift off the holding down studs. If it proves stubborn, it can be tapped free using a hammer and a block of wood, or a hide mallet. Avoid the temptation to lever the joint apart, as this can easily cause damage, and at the very least will mark the surface finish of the casting.

4 **Important note:** Before removing the cylinder block, it should be noted that there is likely to be an accumulation of road dirt around the base of the holding down studs (see photograph). Unless great care is taken, this will drop down into the crankcase during removal, necessitating crankcase separation. Try to arrange the unit so that the block faces downwards, permitting the debris to drop clear of the crankcase mouths. Clean the studs carefully before turning the unit up the right way again.

7.1a Camshaft chain tensioner body is retained by two bolts

7.1b Remove the tachometer drive before camshafts are removed

7.1c Release screws and remove guide sprocket assembly

7.2 Disengage each camshaft from chain, and lift clear

7.4 Take care that road dirt around studs does not enter crankcase

8 Dismantling the engine and gearbox: removing the pistons and piston rings

1 Remove the circlips from the pistons by inserting a screwdriver (or a piece of welding rod chamfered one end), through the groove at the rear of the piston. Discard them. Never re-use old circlips during the rebuild.
2 Using a drift of suitable diameter, tap each gudgeon pin out of position, supporting each piston and connecting rod in turn. Mark each piston inside the skirt so that it is replaced in the appropriate bore. If the gudgeon pins are a tight fit in the piston bosses, it is advisable to warm the pistons. One way is to soak a rag in very hot water, wring the water out and wrap the rag round the piston very quickly. The resultant expansion should ease the grip of the piston bosses on the steel pins.

9 Dismantling the engine and gearbox: removing the contact breaker assembly

1 Remove the contact breaker cover by undoing the two screws that hold the cover on the front right hand side of the engine. Then remove the three screws holding the contact breaker assembly with condensers attached, lift off the plate, separating the wires by pulling them out of the connector. Note that the three backplate screws **only** should be removed. The position of the backplate can be marked in relation to the crankcase as an aid to reassembly, although it will still be necessary to re-time the ignition on reassembly.
2 Holding the crankshaft with a 17 mm spanner on the larger hexagon, slacken the automatic timing unit (ATU) retaining bolt, and pull the unit off the shaft. The unit is located by a roll pin which will remain in position in the crankshaft end.

10 Dismantling the engine and gearbox: removing the clutch and kickstart spring

1 Remove the clutch cover screws, taking care to catch any residual oil which may run out as the cover is lifted away. Slacken the five clutch spring bolts, and remove them together with the washers and clutch springs. Lift away the clutch cover, and withdraw the mushroom-headed pushrod. Tip the unit to displace the small steel ball which fits behind the pushrod.
2 Remove the clutch plates, then obtain a piece of steel strip with which the clutch centre may be spragged to prevent its rotation while the clutch centre nut is slackened. Remove the centre nut and the special washer behind it, followed by the clutch centre, washer and outer drum. If the caged needle roller

bearing remains on the shaft, this should also be slid off.
3 Remove the shims from the end of the kickstart spindle, and pull out the white nylon spring guide. Disengage the spring end using a pair of pointed-nose pliers and allow it to unwind. The spring can now be removed from the shaft.

8.2 Pad crankcase mouths with rag to catch displaced circlips

11 Dismantling the engine and gearbox unit: removing the alternator

1 Remove the screws which secure the left-hand outer cover, and lift the cover and stator assembly away. It will be necessary to prevent the crankshaft from turning while the rotor securing nut is slackened. This can be accomplished by selecting top gear and applying the rear brake if the engine is still in the frame. If the unit is being dismantled for overhaul, a bar can be passed through one of the small end eyes, and supported by small wooden blocks at each side of the crankcase mouth. The nut can now be slackened and removed.
2 In the absence of the official Kawasaki rotor extractor (part number 57001 – 254) a conventional legged puller can be arranged to draw the rotor off its taper. Ensure that the feet of the puller engage squarely on the underside of the rotor, then tighten the centre bolt gradually. If it proves stubborn, a tap on the end of the centre bolt will usually succeed in jarring it free. On no account strike the rotor itself during removal.

9.1 Contact breaker assembly is retained by three screws (arrowed)

11.2 Legged puller may be used to draw rotor off

12 Dismantling the engine and gearbox unit: removing the selector mechanism cover and components

1 Remove the three gearbox sprocket guard bolts and lift the guard away. The selector mechanism cover can be detached after the retaining screws have been removed. The selector shaft and mechanism can now be disengaged and lifted away. The adjacent bearing cap can also be detached and placed to one side. It is retained by two screws.

13 Dismantling the engine and gearbox unit: removing the oil filter, sump and oil pump

1 If the oil filter has not already been removed, the central securing bolt should be released and the cover lifted off, noting that a certain amount of residual oil will be released from the housing. Remove the filter element, and empty and clean out the housing.
2 Slacken and remove the sump mounting bolts, then lift the sump away. It will be noted that the sump is fitted with an oil pressure relief valve, which need not be disturbed at this stage.
3 Slacken the two bolts which retain the oil pump assembly to the inside of the casing, and lift the pump unit away complete with the gauze strainer assembly.

14 Dismantling the engine and gearbox unit: removing the secondary shaft

1 A secondary shaft, incorporating the starter motor freewheel unit, is located below and to the rear of the crankshaft assembly, and is coupled to it by way of a Morse chain. The shaft can be displaced from the casing and withdrawn from the right-hand side (note: left-hand when viewed from the underside of the unit), following the procedure detailed below.
2 Slacken and remove the countersunk screws which secure the retainer plate behind the pinion on the right-hand side of the casing. The plate will drop free as the shaft is displaced. Using a drift, tap the shaft through from the left-hand side, supporting the freewheel and sprocket unit as the shaft is withdrawn. The right-hand bearing will remain in position on the shaft. Disengage the freewheel unit from the Morse chain and place it and the shaft to one side.

12.1a Remove gearbox sprocket guard and selector mechanism

12.1b Disengage and remove the selector mechanism

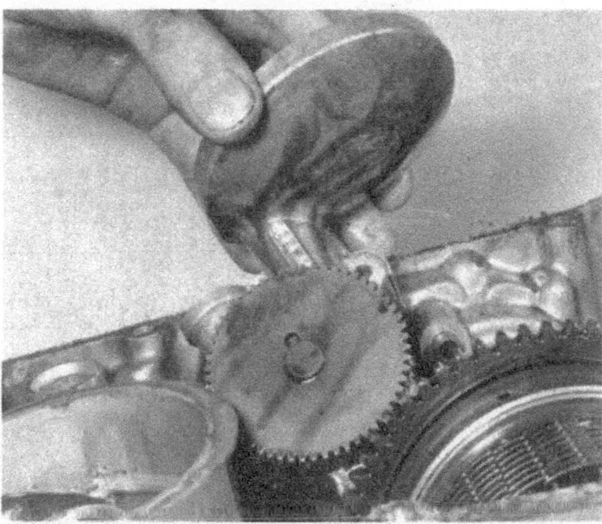
13.3 Oil pump is retained by two bolts

14.3a Release retaining plate screws before shaft is driven out

14.3b Shaft can be removed, leaving ...

14.3c ... secondary shaft sprocket/pinion assembly to be withdrawn

15 Dismantling the engine and gearbox unit: separating the crankcase halves

1 Remove the twelve upper crankcase bolts from the rear half of the casing, then turn the unit over, and remove the eight 6 mm bolts and the ten 8 mm bolts. The casing halves can now be separated.
2 It will probably be found that the two halves are firmly stuck together due to the combined effects of the jointing compound and the dowels. To facilitate separation, leverage points are provided at the jointing face, and a screwdriver can be used here to break the joint. Alternatively, a hammer may be used with a block of wood interposed between it and the casing, to jar the casing halves apart. No hard and fast rule can be given as to which method or methods will prove most successful, and it is largely a matter of discretion on the part of the owner when deciding which to adopt. Usually the joint will break fairly easily, but it may be found that separation will be impaired during the first half inch or so of removal. Check carefully to determine which component or components is sticking, and take steps to release it before proceeding further. Separate the crankcase halves with the unit inverted on the bench, drawing the lower half off the upper half. The crankshaft assembly and gearshafts and clusters will remain in position in the upper half.

16 Dismantling the engine and gearbox unit: removing the upper crankcase half components

1 The crankshaft assembly can be lifted out of the upper casing half, and the camshaft drive chain and Morse secondary shaft drive chain disengaged and removed.
2 The gearbox mainshaft and layshaft can also be lifted out of the casing, noting that these have half ring retainers fitted to the bearing grooves. The camshaft drive chain guide sprocket may be left in position unless there is specific need to remove it.

17 Dismantling the engine and gearbox unit: removing the lower crankcase half components

1 The selector mechanism has three selector forks, two of which are supported by a selector fork shaft. Support the two forks and withdraw the shaft from the left-hand side. The forks can now be lifted out and replaced on the shaft in their correct relative positions. The third gear selector fork fits round the selector drum itself, and has a locating pin which runs in the selector drum track. Remove and discard the split pin which retains the locating pin, then remove the latter using a pair of pointed-nosed pliers.
2 The selector drum is located by a special bolt which screws into the casing and engages in a locating groove. This bolt should be removed after bending back the locking tap. The detent plunger is located at the opposite end of the selector drum, and should also be removed. Detach the large external circlip which retains the detent plate to the end of the selector drum. The plate can now be removed, and the selector drum displaced and withdrawn, leaving the selector fork to be lifted clear of the casing half.
3 The kickstart mechanism, less the spring and guide which have already been removed, is retained by a circlip and a locating plate on the outside face of the casing. Remove the circlip from its groove on the shaft. The locating plate can be removed after releasing the two countersunk retaining screws. The large sleeve in which the shaft runs can now be withdrawn, allowing the mechanism to be removed from the inside of the casing.

15.2 Remove crankcase screws, and lift off lower casing half

16.1 Crankshaft assembly and gearbox clusters can be lifted out

17.1a Withdraw selector fork shaft and remove forks from casing

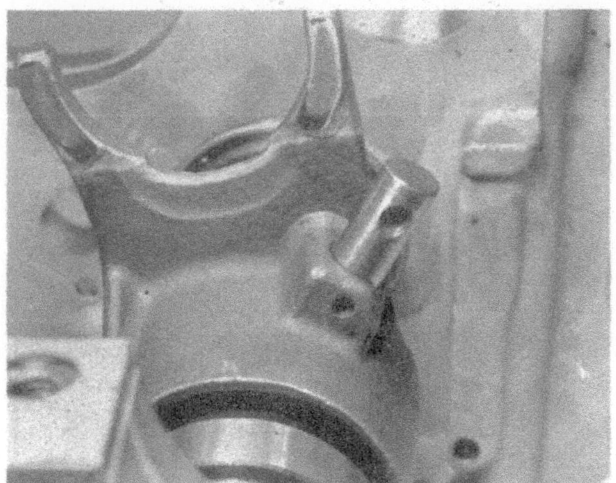
17.1b Remove split pin and locating pin to free selector fork

17.2 Withdraw selector drum and remove fork

17.3a Remove screws, and circlip on shaft, ...

17.3b ... withdraw sleeve from casing, ...

17.3c ... and lift mechanism out of casing half

18 Examination and renovation: general

1 Before examining the parts of the dismantled engine unit for wear, it is essential that they should be cleaned thoroughly. Use a petrol/paraffin mix to remove all traces of old oil and sludge which may have accumulated within the engine.
2 Examine the crankcase castings for cracks or other signs of damage. If a crack is discovered it will require a specialist repair.
3 Examine carefully each part to determine the extent of wear, checking with the tolerance figures listed in the Specifications section of this Chapter. If there is any question of doubt play safe and renew.
4 Use a clean lint free rag for cleaning and drying the various components. This will obviate the risk of small particles obstructing the internal oilways, and causing the lubrication system to fail.

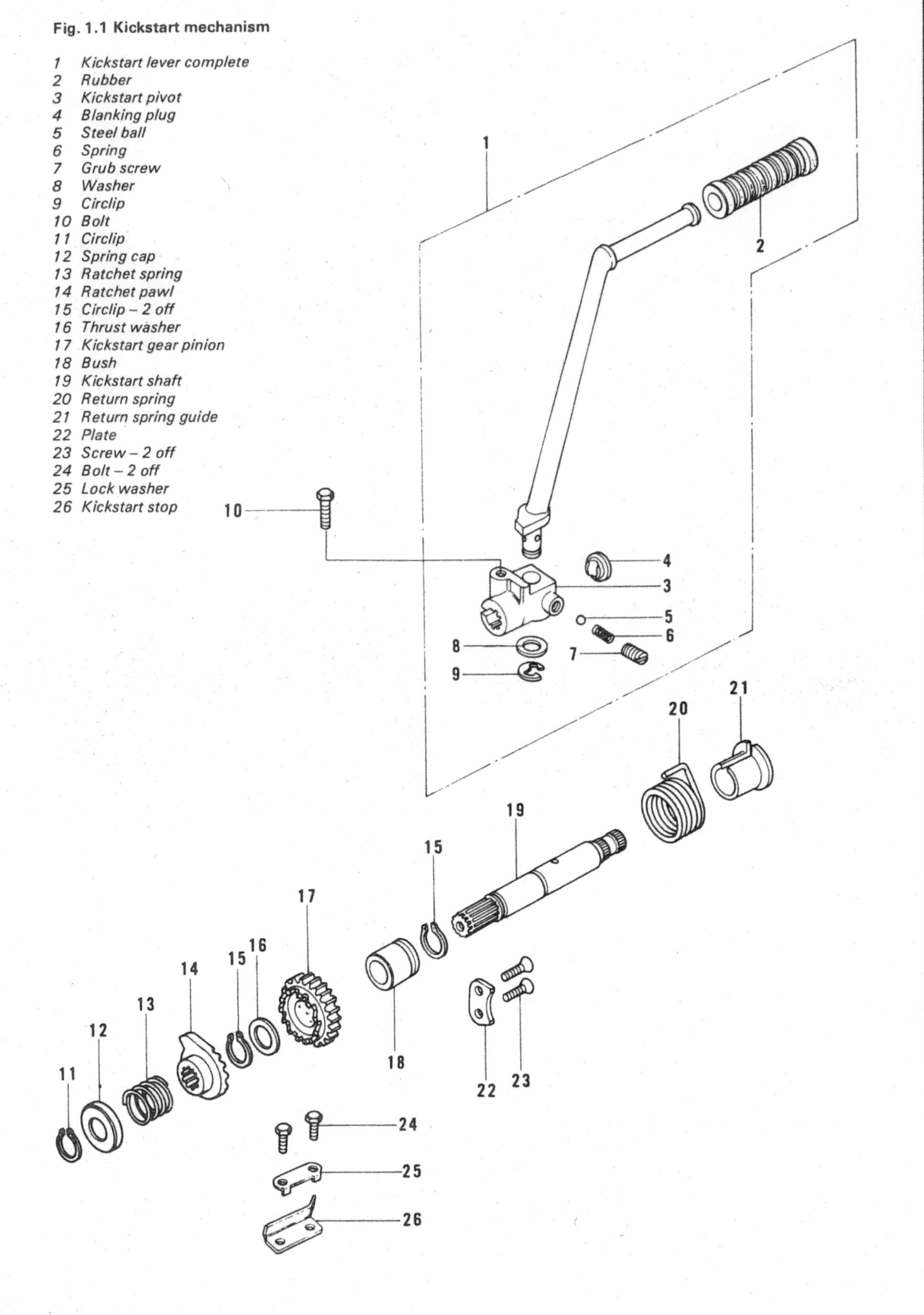

Fig. 1.1 Kickstart mechanism

1 Kickstart lever complete
2 Rubber
3 Kickstart pivot
4 Blanking plug
5 Steel ball
6 Spring
7 Grub screw
8 Washer
9 Circlip
10 Bolt
11 Circlip
12 Spring cap
13 Ratchet spring
14 Ratchet pawl
15 Circlip – 2 off
16 Thrust washer
17 Kickstart gear pinion
18 Bush
19 Kickstart shaft
20 Return spring
21 Return spring guide
22 Plate
23 Screw – 2 off
24 Bolt – 2 off
25 Lock washer
26 Kickstart stop

19 Big end and main bearings: examination and renovation

1 The Kawasaki Z650 models are fitted with shell type bearings on the crankshaft and the big-end assemblies.
2 Bearing shells are relatively inexpensive and it is prudent to renew the entire set of main bearing shells when the engine is dismantled completely, especially in view of the amount of work which will be necessary at a later date if any of the bearings fail. Always renew the five sets of main bearings together.
3 Wear is usually evident in the form of scuffing or score marks in the bearing surface. It is not possible to polish these marks out in view of the very soft nature of the bearing surface and the increased clearance that will result. If wear of this nature is detected, the crankshaft must be checked for ovality as described in the following section.
4 Failure of the big-end bearings is invariably accompanied by a pronounced knock within the crankcase. The knock will become progressively worse and vibration will also be experienced. It is essential that bearing failure is attended to without delay because if the engine is used in this condition there is a risk of breaking a connecting rod or even the crankshaft, causing more extensive damage.
5 Before the big-end bearings can be examined the bearing caps must be removed from each connecting rod. Each cap is retained by two high tensile bolts. Before removal, mark each cap in relation to its connecting rod so that it may be replaced correctly. As with the main bearings, wear will be evident as scuffing or scoring and the bearing shells must be replaced as four complete sets.
6 Replacement bearing shells for either the big-end or main bearings are supplied on a selected fit basis (ie; bearings are selected for correct tolerance to fit the original journal diameter), and it is essential that the parts to be used for renewal are of identical size.
7 Bearing shells should be selected in accordance with the size markings on both the connecting rod and crankshaft. See the following table of sizes:

Connecting rod mark:	1	1	None	None
Crankshaft mark:	1	None	1	None
Use shell type:	B	A	C	B

Shell letter	A	B	C
Part number	PN13034-050	PN13034-051	PN13034-052
Thickness (mm):	1.485-1.490	1.480-1.485	1.475-1.480
Thickness (in):	0.0585-0.0587	0.0583 0.0585	0.0581-0.0583

19.5a Big end bearings should be removed for examination

19.5b Connecting rods, caps and shells are arranged as shown

19.5c Ensure that locating tab engages in cutout

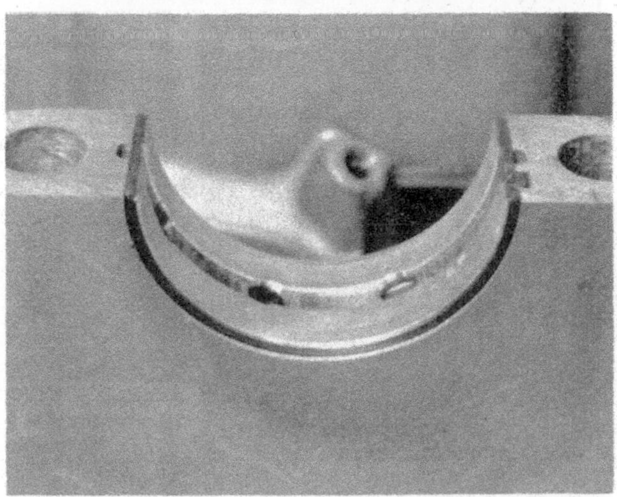
19.5d Check the condition of main bearing shells - note oilways

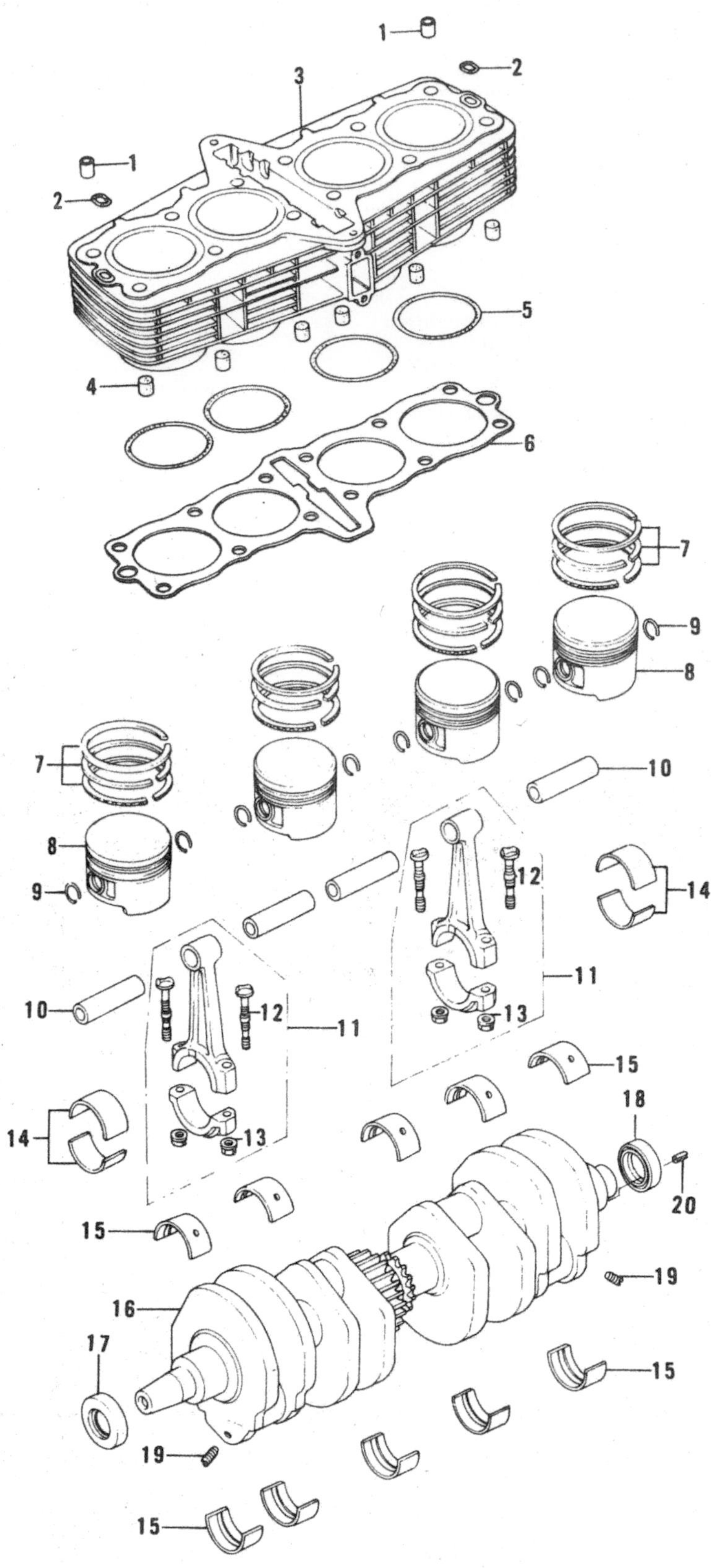

Fig. 1.2 Cylinder block, pistons and crankshaft

1 *Dowel pin – 2 off*
2 *O ring – 2 off*
3 *Cylinder block*
4 *Rubber plug – 6 off*
5 *O ring – 4 off*
6 *Cylinder base gasket*
7 *Piston ring set – 4 off*
8 *Piston – 4 off*
9 *Circlip – 8 off*
10 *Gudgeon pin – 4 off*
11 *Connecting rod assembly – 4 off*
12 *Big end bolt – 8 off*
13 *Big end nut – 8 off*
14 *Big end bearing shell – 8 off*
15 *Main bearing shell – 10 off*
16 *Crankshaft assembly*
17 *Oil seal*
18 *Oil seal*
19 *Grub screw – 2 off*
20 *Dowel pin*

20 Examination and renovation: crankshaft assembly

1 If wear has necessitated the renewal of the big-end and/or main bearing shells, the crankshaft should be checked with a micrometer to verify whether ovality has occurred. If the reading on any one journal varies by more than 0.05 mm (0.002 inch) the crankshaft should be renewed.
2 Mount the crankshaft by supporting both ends on V blocks or between centres on a lathe and check the run-out at the centre main bearing surfaces by means of a dial gauge. The run-out will be half that of the gauge readings indicated. The correct run-out as standard is under 0.02 mm (0.0008 inch) and if it exceeds 0.05 mm (0.002 inch) the crankshaft should be renewed.
3 The clearance between any set of bearings and their respective journal may be checked by the use of Plastigauge (press gauge). Plastigauge is a graduated strip of plastic material that can be compressed between two mating surfaces. The resulting width of the material when measured with a micrometer will give the amount of clearance. For example if the clearance in the big-end bearing is to be measured, Plastigauge should be used in the following manner.

Cut a strip of Plastigauge to the width across the bearing to be measured. Place the Plastigauge strip across the bearing journal so that it is parallel with the crankshaft. Place the connecting rod complete with its half shell on the journal and then carefully replace the bearing cap complete with half shell onto the connecting rod bolts. Replace and tighten the retaining nuts to the correct torque and then loosen and remove the nuts and the bearing cap. Without bending or pressing the Plastigauge strip, place it at its thickest point between a micrometer and read off the measurement. This will indicate the precise clearance. The original size and wear limit of the crankshaft journals and the standard and service limit clearance between all the bearings is given in the specifications at the beginning of this Chapter.
4 The crankshaft has drilled oil passages which allow oil to be fed under pressure to the working surfaces. Care must be taken to clean these out carefully, preferably by using compressed air.
5 When refitting the connecting rods and shell bearings, note that under no circumstances should the shells be adjusted with a shim, 'scraped in' or the fit 'corrected' by filing the connecting rod and bearing cap or by applying emery cloth to the bearing surface. Treatment such as this will end in disaster; if the bearing fit is not good, the parts concerned have not been assembled correctly. This advice also applies to the main bearing shells. Use new big-end bolts too - the originals may have stretched and weakened.
6 Oil the bearing surfaces before reassembly takes place and make sure the tags of the bearing shells are located correctly. After the initial tightening of the connecting rod nuts, check that each connecting rod revolves freely, then tighten to a torque setting of 2.6 - 3.0 kg m (19 - 22 ft lb). Check again that the bearing is quite free.

20.1 Examine crankshaft journals, and condition of sprockets

21 Secondary shaft components: examination and renovation

1 Power from the crankshaft is transmitted by way of a Morse chain to a sprocket on the secondary shaft, which in turn drives the clutch. The secondary shaft also incorporates a rubber-segment shock absorber which counteracts any engine vibration. It is not normally necessary to dismantle the secondary shaft components unless one of the following symptoms has been apparent:

a) Starter motor not engaging or disengaging correctly, indicating wear in the clutch rollers, weak or broken springs or damaged clutch housing.
b) Snatch in primary transmission, indicating wear or damage to the shock absorber rubbers or hub.

2 If starter clutch problems are in evidence, dismantle the unit by sliding the starter gear off the end of the shaft, together with its needle roller bearing. The clutch body will now be exposed, and will be seen to contain three sets of springs, caps and rollers. These should be removed and examined for wear or damage. Although this unit is not especially prone to wear, look for signs of flats appearing on the roller faces, and for signs of wear in the clutch body and on the gear boss on which the rollers act. Wear in these areas can cause the clutch to jam, and prevent the starter motor from disengaging correctly. Conversely, it can also cause slipping, preventing the drive from the starter motor from being transmitted to the engine. The only safe course of action, if wear is evident, is to renew the parts concerned. If the rollers are to be renewed, fit new springs as a matter of course, to avoid subsequent problems in the event of their failure.
3 Like the starter clutch, the shock absorber components rarely give any trouble. The unit is, however, very easy to dismantle. The main body, which incorporates the Morse drive sprocket, is retained by a circlip. When this has been released, the body can be slid off and the rubber segments removed for examination. Any damage will be self-evident and normally will be confined to the rubber segments. These will tend to become compressed and rounded off after a very high mileage, and should be renewed if this is the case.
4 Examine the teeth on the outside of the housing, looking for chips and signs of wear. If the teeth are only slightly marked, they may be reclaimed, using a fine oilstone. More severe damage will necessitate renewal.
5 The journal ball bearings which support the shaft should be checked for signs of roughness and free play after they have been washed in clean petrol and dried off. Any sign of grittiness or slop is indicative of the need for renewal. The bearing which is still in place in the casing can be driven out using a large diameter socket as a drift. The remaining bearing can be pulled off the shaft by way of a bearing extractor or small sprocket puller.
6 The Morse primary chain has no provision for adjustment, but will normally cover a very high mileage before renewal becomes necessary. Wear can be checked by temporarily re-installing the crankshaft and secondary shaft in the casing half, with the chain fitted in its normal position. Free play should be measured at the middle of the run, and should not exceed 27 mm (1.063 in). If worn beyond this amount, a new chain must be fitted.

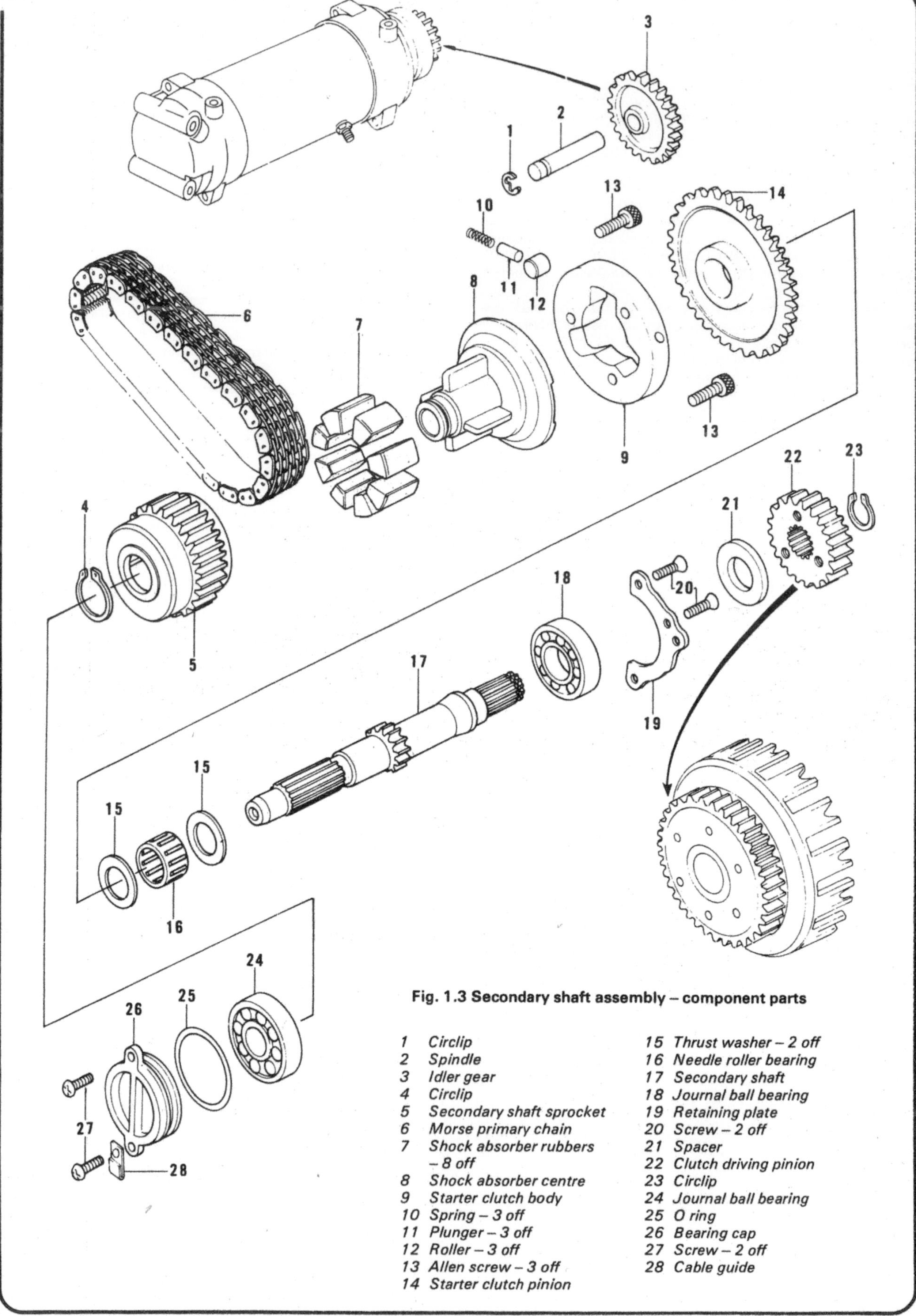

Fig. 1.3 Secondary shaft assembly – component parts

1 *Circlip*
2 *Spindle*
3 *Idler gear*
4 *Circlip*
5 *Secondary shaft sprocket*
6 *Morse primary chain*
7 *Shock absorber rubbers – 8 off*
8 *Shock absorber centre*
9 *Starter clutch body*
10 *Spring – 3 off*
11 *Plunger – 3 off*
12 *Roller – 3 off*
13 *Allen screw – 3 off*
14 *Starter clutch pinion*
15 *Thrust washer – 2 off*
16 *Needle roller bearing*
17 *Secondary shaft*
18 *Journal ball bearing*
19 *Retaining plate*
20 *Screw – 2 off*
21 *Spacer*
22 *Clutch driving pinion*
23 *Circlip*
24 *Journal ball bearing*
25 *O ring*
26 *Bearing cap*
27 *Screw – 2 off*
28 *Cable guide*

21.2a Pinion can be lifted away from clutch assembly

21.2b Check rollers for wear or flats, ensure that springs work

22 Oil seals: examination and replacement

1 Oil seal failure is difficult to define precisely. Usually it takes the form of oil showing on the outside of the machine, and there is nothing worse than those unsightly patches of oil on the ground where the machine has been standing. One of the most crucial places to look for an oil leak is behind the gearbox final drive sprocket. The seal and 'O' ring that fits on the shaft should be renewed if there is any sign of a leak.
2 Oil seals are relatively inexpensive, and if the unit is being overhauled it is advisable to renew all the seals as a matter of course. This will preclude any risk of an annoying oil leak developing after the unit has been reinstalled in the frame.

23 Cylinder block: examination and renovation

1 The usual indication of badly worn cylinder bores and pistons is excessive smoking from the exhausts. This usually takes the form of blue haze tending to develop into a white haze as the wear becomes more pronounced.
2 The other indication is piston slap, a form of metallic rattle which occurs when there is little load on the engine. If the top of the bore is examined carefully, it will be found that there is a ridge on the thrust side, the depth of which will vary according to the rate of wear which has taken place. This marks the limit of travel of the top piston ring.
3 Measure the bore diameter just below the ridge using an internal micrometer,or a dial gauge. Compare the reading you obtain with the reading at the bottom of the cylinder bore, which has not been subjected to any piston wear. If the difference in readings exceeds 0.05 mm (0.002 in) the cylinder block will have to be bored and honed, and fitted with the required oversize pistons.
4 If a measuring instrument is not available, the amount of cylinder bore wear can be measured by inserting the piston (without rings) so that it is approximately $\frac{3}{4}$ inch from the top of the bore. If it is possible to insert a 0.005 inch feeler gauge between the piston and cylinder wall on the thrust side of the piston, remedial action must be taken.
5 Kawasaki supply pistons in two oversizes: 0.5 mm (0.020 unch) and 1.0 mm (0.040 inch). If boring in excess of 1.0 mm becomes necessary, the cylinder block must be renewed since new liners are not available from Kawasaki.
6 Make sure the external cooling fins of the cylinder block are free from oil and road dirt, as this can prevent the free flow of air over the engine and cause overheating problems.

24 Pistons and piston rings: examination and renovation

1 If a rebore becomes necessary, the existing pistons and piston rings can be disregarded because they will have to be replaced by their new oversizes.
2 Remove all traces of carbon from the piston crowns, using a blunt ended scraper to avoid scratching the surface. Finish off by polishing the crowns of each piston with metal polish, so that carbon will not adhere so rapidly in the future. Never use emery cloth on the soft aluminium.
3 Piston wear usually occurs at the skirt or lower end of the piston and takes the form of vertical streaks or score marks on the thrust side of the piston. Damage of this nature will necessitate renewal.
4 The piston ring grooves may become enlarged in use, allowing the rings to have a greater side float. If the clearance exceeds 0.15 mm (0.006 in) the pistons are due for replacement.
5 To measure the end gap, insert each piston ring into its cylinder bore, using the crown of the bare piston to locate it about 1 inch from the top of the bore. Make sure it is square in the bore and insert a feeler gauge in the end gap of the ring. If the end gap exceeds 0.7 mm (0.028 inch) the ring must be renewed. The standard gap is 0.15 – 0.3 mm (0.006 – 0.012 in).

When refitting new piston rings, it is also necessary to check the end gap. If there is insufficient clearance, the rings will break up in the bore whilst the engine is running and cause extensive damage. The ring gap may be increased by filing the ends of the rings with a fine file.

The ring should be supported on the end as much as possible to avoid breakage when filing, and should be filed square with the end. Remove only a small amount of metal at a time and keep rechecking the clearance in the bore.

25 Cylinder head: examination and renovation

1 Remove all traces of carbon from the cylinder head using a blunt ended scraper (the round end of an old steel rule will do). Finish by polishing with metal polish to give a smooth shiny surface. This will aid gas flow and will also prevent carbon from adhering so firmly in the future.
2 Check the condition of the sparking plug hole threads. If the threads are worn or crossed they can be reclaimed by a Helicoil insert. Most motorcycle dealers operate this service which is very simple, cheap and effective.

3 Clean the cylinder head fins with a wire brush, to prevent overheating, through dirt blocking the fins.
4 Lay the cylinder head on a sheet of $\frac{1}{4}$ inch plate glass to check for distortion. Aluminium alloy cylinder heads distort very easily, especially if the cylinder head bolts are tightened down unevenly. If the amount of distortion is only slight, it is permissible to rub the head down until it is flat once again by wrapping a sheet of very fine emery cloth around the plate glass base and rubbing with a rotary motion.
5 If the cylinder head is distorted badly (one way of determining this is if the cylinder head gaskets have a tendency to keep blowing), the head will have to be machined by a competent engineer experienced in this type of work. This will, of course, raise the compression of the engine, and if too much is removed can adversely affect the performance of the engine. If there is risk of this happening, the only remedy is a new replacement cylinder head.

26 Valves, valve seats, and valve guides: examination and renovation

1 Remove the valve tappets and shims, keeping them separate for installation in their original locations. Compress the valve springs with a valve spring compressor, and remove the split valve collets, also the oil seals from the valve guides, as it is best to renew these latter components.
2 Remove the valves and springs, making sure to keep to the locations during assembly. Inspect the valves for wear, overheating or burning, and replace them as necessary. Normally, the exhaust valves will need renewal far more often than the inlet valves, as the latter run at relatively low temperatures. If any of the valve seating faces are badly pitted, do not attempt to cure this by grinding them, as this will invariably cause the valve seats to become pocketed. It is permissible to have the valve(s) refaced by a motorcycle specialist or small engineering works. The valve seating angle is 45°. The valve must be renewed if the head thickness (the area between the edge of the seating surface and the top of the head) is reduced to 0.5 mm (0.020 in). The nominal thickness is 1.0 mm (0.040 in).
3 Measure the bore of each valve guide in at least four places using a small bore gauge and micrometer. The standard measurement for each guide (internal diameter) is 7.000 – 7.015 mm (0.2756 – 0.2761 in). If the measurement exceeds 7.08 mm (0.2787 in) the guide should be replaced with a new one.

If a small bore gauge and micrometer are not available, insert a new valve into the guide, and set a dial gauge against the valve stem. Gently move the valve back and forth in the guide and measure the travel of the valve in each direction. The guide will have to be renewed if the clearance between the valve and guide exceeds the following figures:

	Nominal	*Wear limit*
Inlet	0.049 – 0.085 mm (0.0019 – 0.0033 in)	0.24 mm (0.0095 in)
Exhaust	0.057 – 0.124 mm (0.0022 – 0.0049 in)	0.19 mm (0.0075 in)

(Note that the above method does not give the actual valve to valve guide clearance).

4 It is worthwhile pausing at this juncture to consider the best course of action. It must be borne in mind that valve guide renewal is not easy, and will require that the valve seats be recut after the guide has been fitted and reamed. It is also remarkably easy to damage the cylinder head unless great care is taken during these operations. It may, therefore, be considered better to entrust these jobs to a competent engineering company or to a Kawasaki Service Agent. For the more intrepid, skilled and better equipped owner, the procedure is as follows:
5 Heat the cylinder head slowly and evenly, in an oven to prevent warpage, to 120 – 150°C (248 – 302°F). Using a stepped drift, tap the guide(s) lightly out of the head, taking care not to burn yourself on the hot casting. New guides should be fitted in a similar manner, ensuring that they seat squarely in the head casting. If a valve guide is loose in the head, it may be possible to have an oversize guide machined and fitted by a competent engineering works, noting that the cylinder head must be bored to suit the new guide. The popular 'dodge' of knurling the outside of the guide is crude and is not recommended.
6 After the guide has been fitted it must be reamed using a Kawasaki reamer (Part Number 57001 – 162).Make sure that the reamer passes squarely though the valve guide bore, taking care not to accidentally gouge out too much material. The valve seat must now be re-cut in the following manner:
7 If a valve guide has been renewed, or a valve seat face is worn or pitted, it must be re-cut to ensure efficient sealing. The process requires the use of three cutters (30°, 45° and 60°). These are normally available as a set. Assemble the tool according to the manufacturer's instructions, with the 45° cutter fitted. Arrange the tool with the spigot inserted in the valve guide, and remove just enough metal to ensure an even, pit-free seating surface. Note that if too much metal is removed, the valve will become pocketed, and the complete cylinder head will have to be renewed.
Kawasaki do not supply valve seat inserts, so the utmost care must be taken.
8 The 30° and 60° cutters should be used next, and in that order, to leave the raised 45° seating face as an even band between 0.5 and 1.0 mm in width. The valve(s) should now be ground-in, in the normal manner.
9 The valves should be ground in, using ordinary oil-bound grinding paste, to remove any pitting or to finish off a newly cut seat. Note that it is not normally essential to resort to using the coarse grade of paste which is normally supplied in dual-grade containers.

Valve grinding is a simple task. Commence by smearing a trace of fine valve grinding compound (carborundum paste) on the valve seat and apply a suction tool to the head of the valve. Oil the valve stem and insert the valve in the guide so that the two surfaces to be ground in make contact with one another. With a semi-rotary motion, grind in the valve head to the seat, using a backward and forward action. Lift the valve occasionally so that the grinding compound is distributed evenly. Repeat the application until an unbroken ring of light grey matt finish is obtained on both valve and seat. This denotes the grinding operation is now complete. Before passing to the next valve, make sure that all traces of the valve grinding compound have been removed from both the valve and its seat and that none has entered the valve guide. If this precaution is not observed, rapid wear will take place due to the highly abrasive nature of the carborundum paste.
10 In view of the number of valves used in these engines, it may be thought worthwhile purchasing one of the oscillatory valve lapping tools which have come onto the market in recent years. This device consists of a sealed gearbox having a driving spindle on one side and a rubber sucker on the other. Rotary movement from an electric drill chuck is converted to the correct to-and-fro motion at the sucker. These devices are well worth having if more than one or two valves are to be lapped. **On no account** fit the valve stem straight into a drill chuck and attempt grinding by that method, as this will quickly destroy the seat.
11 Reassemble the valve and valve springs by reversing the dismantling procedure. Fit new oil seals to each valve guide and oil both the valve stem and the valve guide, prior to reassembly. Take special care to ensure the valve guide oil seal is not damaged when the valve is inserted. As a final check after assembly, give the end of each valve stem a light tap with a hammer, to make sure the split collets have located correctly.

24.2 Clean all traces of carbon from piston crowns

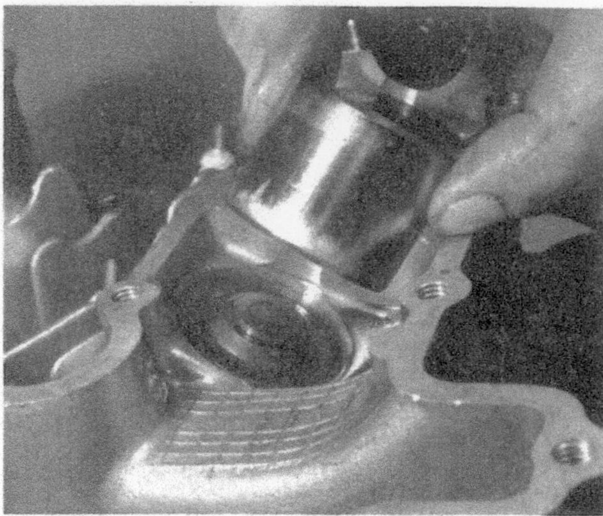
26.1a Lift off tappets and remove shims, keeping them in order

26.1b Use valve spring compressor to allow collets to be removed

26.2 Springs and valves may be removed for examination

26.3 Check guides for wear. Renew oil seals

26.11 Reassemble valves after they have been ground in

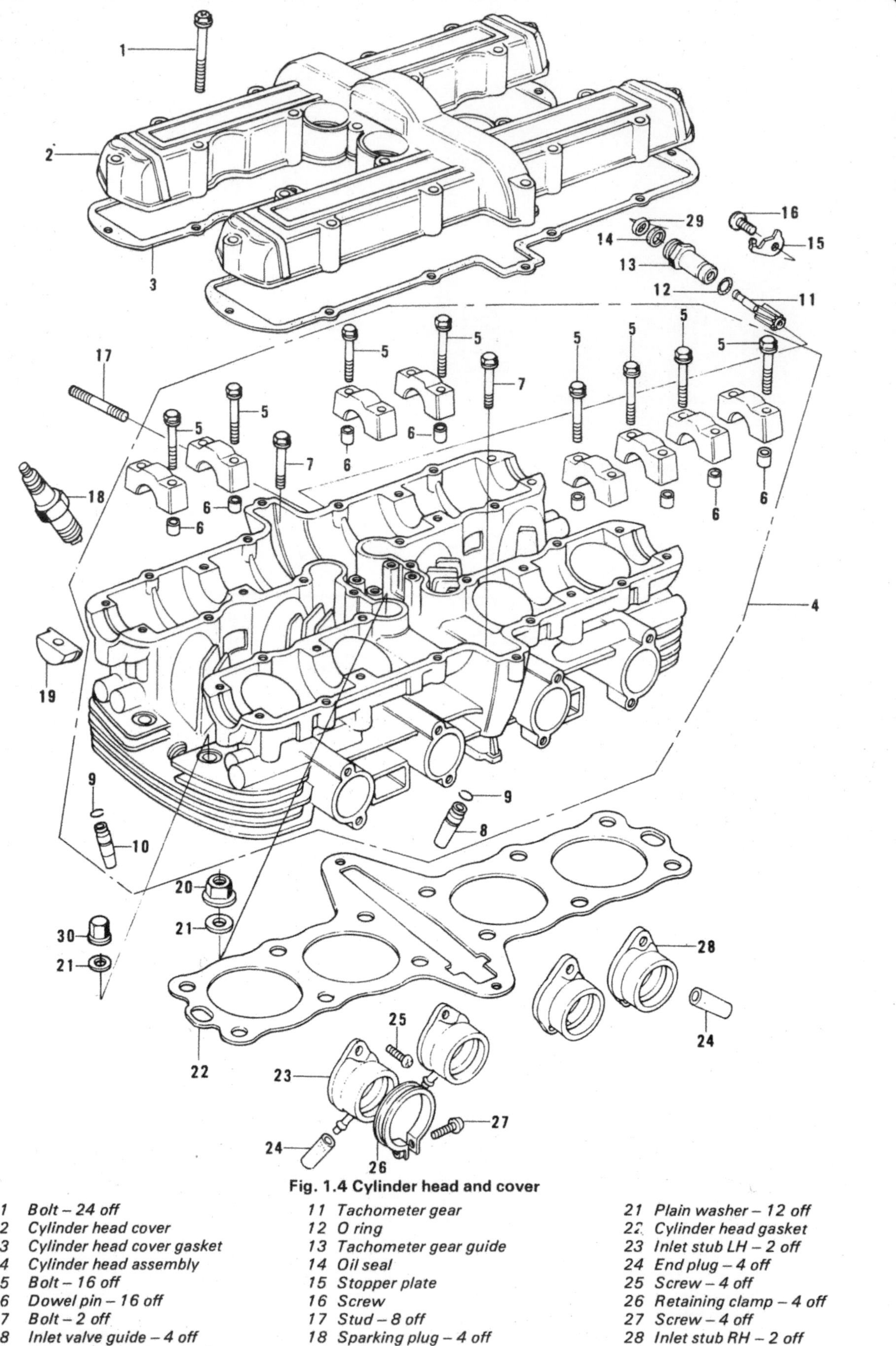

Fig. 1.4 Cylinder head and cover

1 Bolt – 24 off
2 Cylinder head cover
3 Cylinder head cover gasket
4 Cylinder head assembly
5 Bolt – 16 off
6 Dowel pin – 16 off
7 Bolt – 2 off
8 Inlet valve guide – 4 off
9 Valve guide circlip – 8 off
10 Exhaust valve guide – 4 off
11 Tachometer gear
12 O ring
13 Tachometer gear guide
14 Oil seal
15 Stopper plate
16 Screw
17 Stud – 8 off
18 Sparking plug – 4 off
19 Cylinder head cover plug – 4 off
20 Cylinder head nut – 8 off
21 Plain washer – 12 off
22 Cylinder head gasket
23 Inlet stub LH – 2 off
24 End plug – 4 off
25 Screw – 4 off
26 Retaining clamp – 4 off
27 Screw – 4 off
28 Inlet stub RH – 2 off
29 Sealing ring
30 Cylinder head nut – 4 off

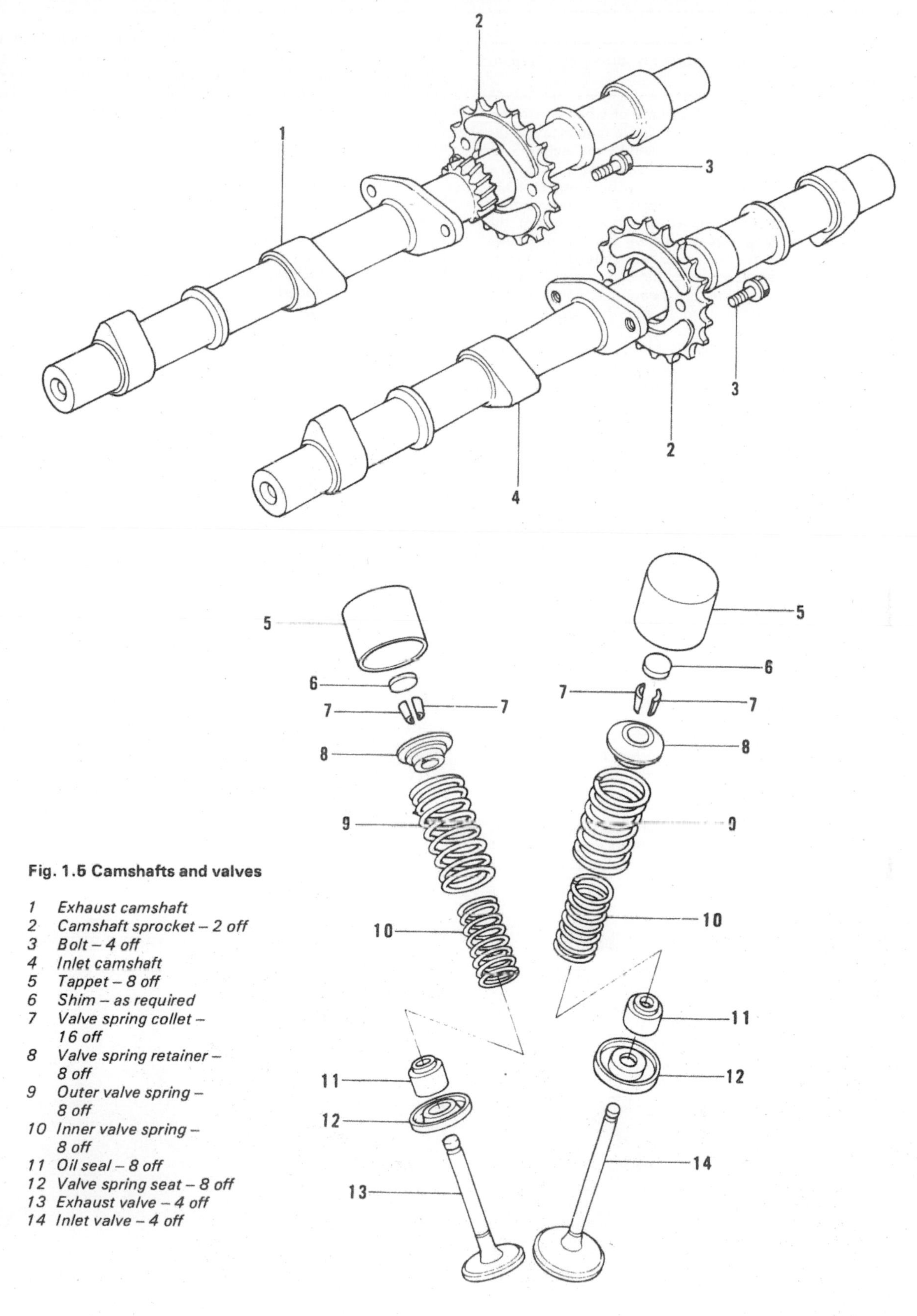

Fig. 1.5 Camshafts and valves

1 *Exhaust camshaft*
2 *Camshaft sprocket – 2 off*
3 *Bolt – 4 off*
4 *Inlet camshaft*
5 *Tappet – 8 off*
6 *Shim – as required*
7 *Valve spring collet – 16 off*
8 *Valve spring retainer – 8 off*
9 *Outer valve spring – 8 off*
10 *Inner valve spring – 8 off*
11 *Oil seal – 8 off*
12 *Valve spring seat – 8 off*
13 *Exhaust valve – 4 off*
14 *Inlet valve – 4 off*

27 Camshafts, tappets and camshaft drive mechanism: examination and renovation

1 Examine the camshaft lobes for signs of wear or scoring. Wear is normally evident in the form of visible flats worn on the peak of the lobes, and this may be checked by measuring each lobe at its widest point. The standard measurement is 35.73 – 35.87 mm (1.4067 – 1.4122 in). If worn to 35.65 mm (1.4035 in) or less the camshaft must be renewed. Scoring or similar damage can usually be attributed to a partial failure of the lubrication system, possibly due to the oil filter element not having been renewed at the specified mileage, causing unfiltered oil to be circulated by way of the bypass valve. Before fitting new camshafts, examine the bearing surfaces of the camshafts, and cylinder head, and rectify the cause of the failure.

2 If the camshaft bearing surfaces are marred, it is likely that renewal of both the cylinder head and the camshafts will be the only solution. This is because the camshafts run directly in the cylinder head casting, using the alloy as a bearing surface. Assemble the bearing caps and measure the internal bore using a bore micrometer. The nominal size is 22.000 – 22.021 mm (0.8661 – 0.8670 in). The cylinder head and caps must be renewed if this figure exceeds 22.060 mm (0.8690 in).

3 Measure the camshaft bearing journals, which when new should be between 21.940 – 21.960 mm (0.8640 – 0.8646 in). Renew the camshaft if worn down to 21.900 mm (0.8622 in) or less. The nominal bearing clearance should be 0.040 – 0.081 mm (0.0016 – 0.0032 in), the wear limit being 0.170 mm (0.0067 in). This can be checked by using the Plastigauge method as described in Section 20 of this Chapter.

4 Camshaft run-out can be checked by supporting each end of the shaft on V-blocks, and measuring any run-out using a dial test indicator running on the camshaft sprocket boss (having first removed the sprocket). This should not normally be more than 0.01 mm (0.004 in). The camshaft must be renewed if run-out exceeds 0.1 mm (0.0039 in).

5 The single camshaft drive chain should be checked for wear, particularly if tensioner adjustment has failed to prevent chain noise, this latter condition being indicative that the chain is probably due for renewal. Lay the chain on a flat surface, and get an assistant to stretch it taut. Using a vernier caliper gauge, measure a 20 link run of the chain. Repeat this check in one or two other places. The nominal length is 160.00 mm (6.299 in). If this exceeds 162.4 mm (6.394 in), the chain must be renewed.

6 The various guide sprockets, the roller guide and the tensioner assembly should be examined for wear or damage, which will normally be fairly obvious. Renew any parts which appear worn or are damaged, especially if a new chain has been fitted. The same can be applied to the two camshaft sprockets. These components can normally be expected to give many miles of service if correctly maintained. The tachometer worm drive on the exhaust camshaft is unlikely to suffer undue wear.

7 The worm drive to the tachometer is an integral part of the camshaft which meshes with a pinion attached to the cylinder head. If the worm is damaged or badly worn, it will be necessary to renew the camshaft complete.

8 The tachometer driven worm gear shaft is fitted in a housing which is a press fit in the cylinder head cover. If the worm gear is chipped or broken the gear and integral shaft should be renewed.

27.6a Examine the tensioner sprocket, ...

27.6b ... guide sprockets, ...

27.6c ... and rollers for wear or damage

27.6d Check the condition of the upper guide sprocket assembly, ...

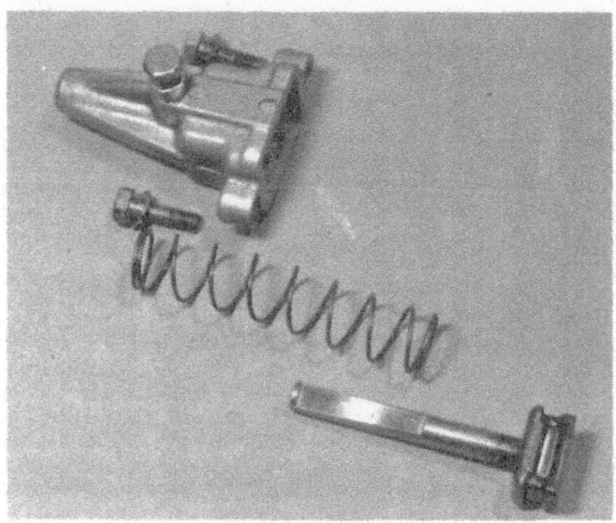

27.6e ... and the tensioner body mechanism

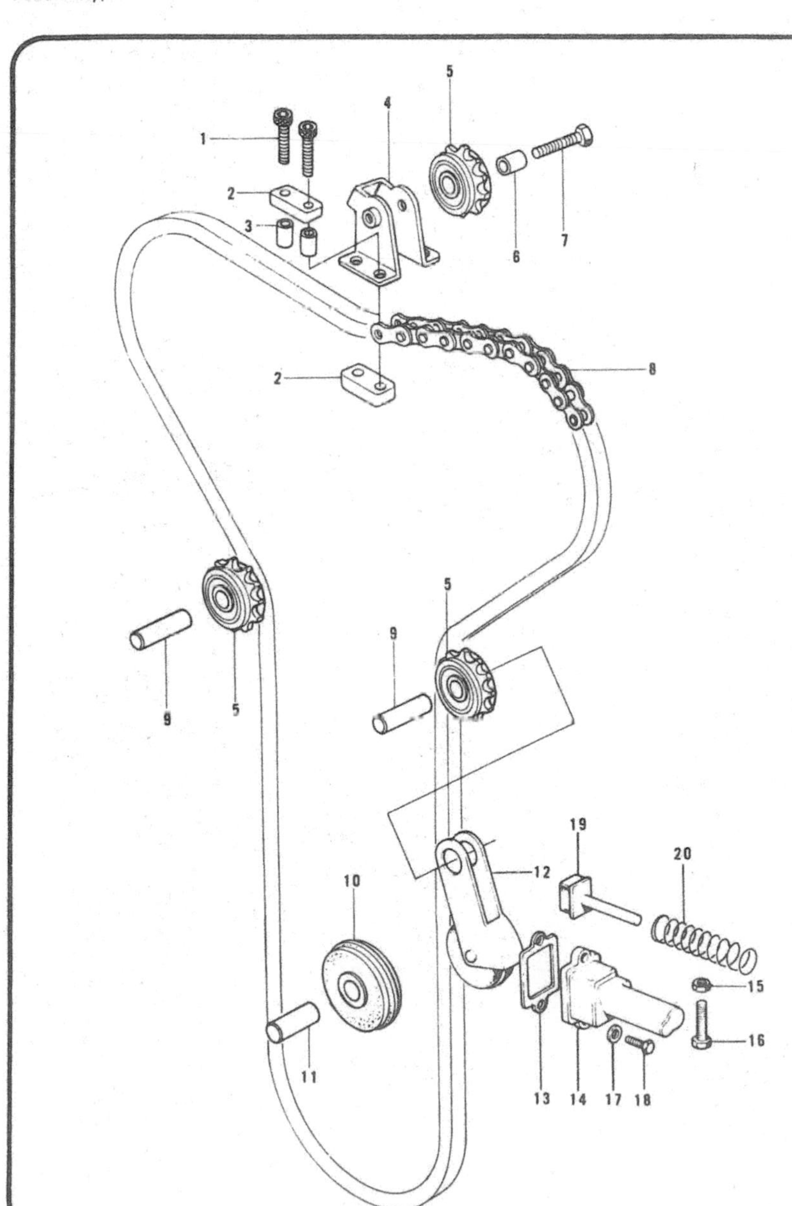

Fig. 1.6 Cam chain and tensioner

1 *Allen screw – 4 off*
2 *Rubber block – 4 off*
3 *Collar – 4 off*
4 *Cam chain upper guide bracket*
5 *Cam chain upper guide sprocket – 3 off*
6 *Collar*
7 *Bolt*
8 *Cam chain*
9 *Idler sprocket shaft – 2 off*
10 *Roller*
11 *Guide roller shaft*
12 *Cam chain tensioner assembly*
13 *Gasket*
14 *Tensioner body*
15 *Nut*
16 *Tensioner adjusting bolt*
17 *Plain washer – 2 off*
18 *Bolt – 2 off*
19 *Pushrod assembly*
20 *Tensioner spring*

28 Clutch: examination and renovation

1 After an extended period of service the clutch linings will wear and promote clutch slip. The limit of wear measured across each inserted plate and the standard measurement is as follows:

	Standard	Service limit
Clutch plate thickness	3.7 – 3.9 mm (0.146 – 0.154 in)	3.5 mm (0.138 in)

When the overall width reaches the limit, the inserted plates must be renewed, preferably as a complete set.

2 The plain plates should not show any excess heating (blueing). Check the warpage of each plate using plate glass or surface plate and a feeler gauge. The maximum allowable warpage is 0.40 mm (0.0157 in) in the case of the plain plates, and 0.30 mm (0.0118 in) in the case of the friction plates.

3 Examine the clutch assembly for burrs or indentation on the edges of the protruding tongues of the inserted plates and/or slots worn in the edges of the outer drum with which they engage. Similar wear can occur between the inner tongues of the plain clutch plates and the slots in the clutch inner drum. Wear of this nature will cause clutch drag and slow disengagement during gear changes, since the plates will become trapped and will not free fully when the clutch is withdrawn. A small amount of wear can be corrected by dressing with a fine file; more extensive wear will necessitate renewal of the worn parts.

Note that the clearance between the clutch drum slots and the tangs of the clutch plates must not exceed 1.0 mm (0.040 in).

4 The clutch release mechanism attached to the final drive sprocket cover does not normally require attention provided it is greased at regular intervals. It is held to the cover by two cross-head screws and operates on the worm and quick start thread principle.

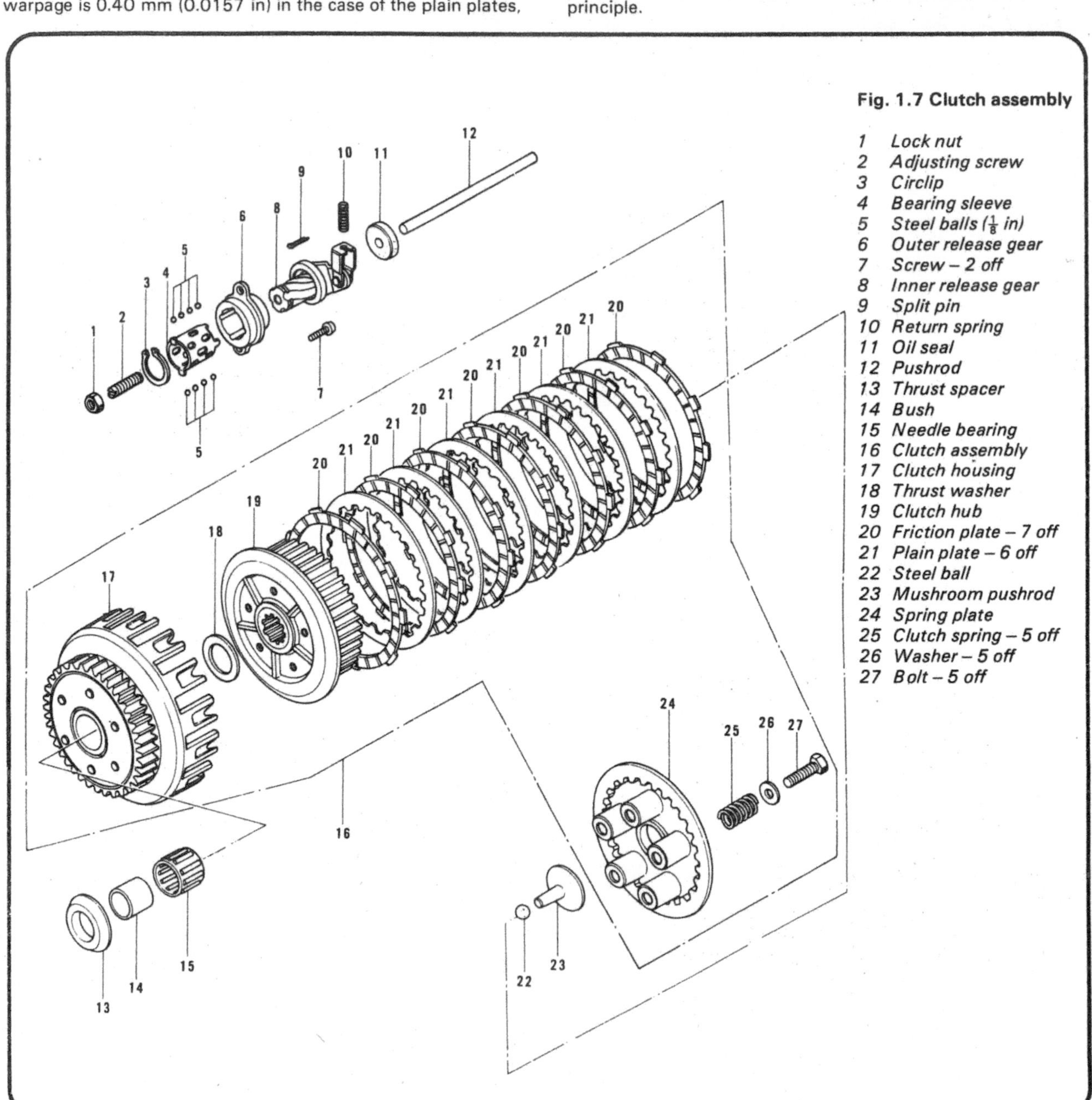

Fig. 1.7 Clutch assembly

1 *Lock nut*
2 *Adjusting screw*
3 *Circlip*
4 *Bearing sleeve*
5 *Steel balls ($\frac{1}{8}$ in)*
6 *Outer release gear*
7 *Screw – 2 off*
8 *Inner release gear*
9 *Split pin*
10 *Return spring*
11 *Oil seal*
12 *Pushrod*
13 *Thrust spacer*
14 *Bush*
15 *Needle bearing*
16 *Clutch assembly*
17 *Clutch housing*
18 *Thrust washer*
19 *Clutch hub*
20 *Friction plate – 7 off*
21 *Plain plate – 6 off*
22 *Steel ball*
23 *Mushroom pushrod*
24 *Spring plate*
25 *Clutch spring – 5 off*
26 *Washer – 5 off*
27 *Bolt – 5 off*

29 Examination and renovation: gearbox components

1 Examine each of the gear pinions to ensure that there are no chipped or broken teeth and that the dogs on the end of the pinions are not rounded. Gear pinions with any of these defects must be renewed; there is no satisfactory method of reclaiming them.
2 After thorough washing in petrol, the bearings should be examined for roughness and play. Also check for pitting on the roller tracks.
3 It is advisable to renew the gearbox oil-seals irrespective of their condition. Should a re-used oil seal fail at a later date, a considerable amount of work is involved to gain access to renew it.
4 Check the gear selector rod for straightness by rolling it on a sheet of plate glass. A bent rod will cause difficulty in selecting gears and will make the gear change particularly heavy.
5 The selector forks should be examined closely, to ensure that they are not bent or badly worn. The case-hardened pegs which engage with the cam channels are easily renewable if they are worn. Under normal conditions, the gear selector mechanism is unlikely to wear quickly, unless the gearbox oil level has been allowed to become low.
6 The tracks in the selector drum, with which the selector forks engage, should not show any undue signs of wear unless neglect has led to under-lubrication of the gearbox. Check the tension of the gearchange pawl, gearchange arm and drum stopper arm springs. Weakness in the springs will lead to imprecise gear selection. Check the condition of the gear stopper arm roller and the pins in the change drum end with which it engages. It is unlikely that wear will take place here except after considerable mileage.
7 Check the condition of the kickstart components. If slipping has been encountered a worn ratchet and pawl will invariably be traced as the cause. Any other damage or wear to the components will be self-evident. If either the ratchet or pawl is found to be faulty, both components must be replaced as a pair. Examine the kickstart return spring, which should be renewed if there is any doubt about its condition.
8 If it is found necessary to renew any of the gearbox components or those of the kickstart mechanism, the accompanying line drawings will give details of the assembly sequence. In addition, gear cluster reassembly is covered in the accompanying photographic sequence.

29.7a Renew ratchet components if badly worn or damaged

27.7b Ensure that reference marks align before refitting circlip

29.8a Fit layshaft 2nd gear, washer and circlip, ...

29.8b ... followed by 5th gear pinion as shown

29.8c Fit circlip and washer, then slide on 3rd gear pinion

29.8d Fit another washer and circlip, followed by 4th gear pinion

29.8e 1st gear pinion has bushed centre ...

29.8f ... and is next component to be fitted

29.8g Finally, fit needle roller bearing, circlip and outer race

29.8h Mainshaft 4th gear pinion also has bushed centre ...

29.8i ... and is the first component to be fitted to shaft

29.8j Locate with washer and circlip, ...

29.8k ... then fit the mainshaft 3rd gear pinion

29.8l Fit circlip to groove as shown, then fit washer ...

29.8m ... and splined bush. This bush carries ...

29.8n ... the mainshaft 5th gear pinion, ...

29.8o ... which is followed by mainshaft 2nd gear pinion

29.8p Finish off with bearing, circlip and outer race

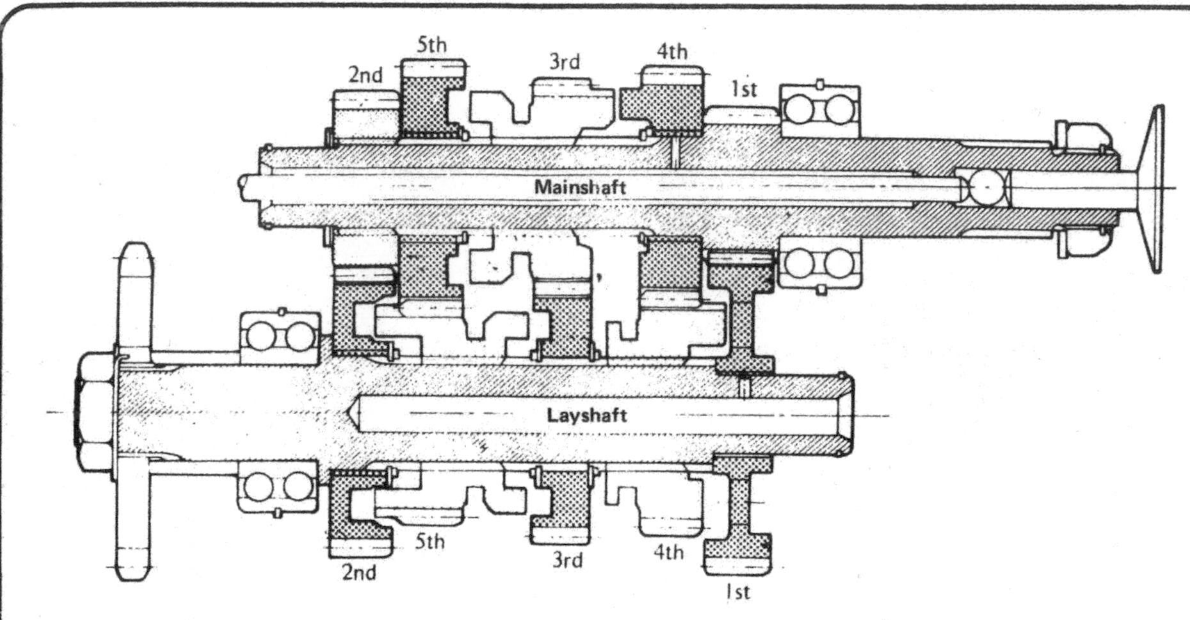

Fig. 1.8 Gearbox: cross-section showing gear positions

Fig. 1.9 Gearbox components (see opposite page)

1 Bush
2 O ring
3 Circlip – 2 off
4 Needle roller bearing – 2 off
5 Steel washer – 2 off
6 Phosphor bronze washer – 2 off
7 Mainshaft 2nd gear pinion
8 Mainshaft top gear pinion
9 Mainshaft top gear bush
10 Washer – 2 off
11 Circlip – 2 off
12 Mainshaft 3rd gear pinion
13 Mainshaft 4th gear pinion
14 Mainshaft
15 Ball bearing – 2 off
16 Washer
17 Lock nut
18 Neutral gear indicator switch
19 Gear change drum
20 Screw
21 Lock washer
22 Pin retainer plate
23 Change drum pin – 5 off
24 Change drum pin
25 Circlip
26 Needle roller bearing
27 4th & top gear selector fork
28 Dowel pin
29 Split pin
30 Cam plate
31 Drive pin
32 Circlip
33 Detent plunger
34 Detent spring
35 Detent body
36 Tab washer
37 Bolt
38 Circlip
39 Selector fork rod
40 2nd & 3rd gear selector fork
41 1st gear selector fork
42 Nut
43 Tab washer
44 Engine sprocket
45 Engine sprocket collar
46 O ring
47 Oil seal
48 Layshaft
49 Layshaft 2nd gear pinion
50 Lock washer
51 Circlip
52 Layshaft top gear pinion
53 Layshaft 3rd gear pinion
54 Layshaft 4th gear pinion
55 Layshaft 1st gear pinion
56 Bush

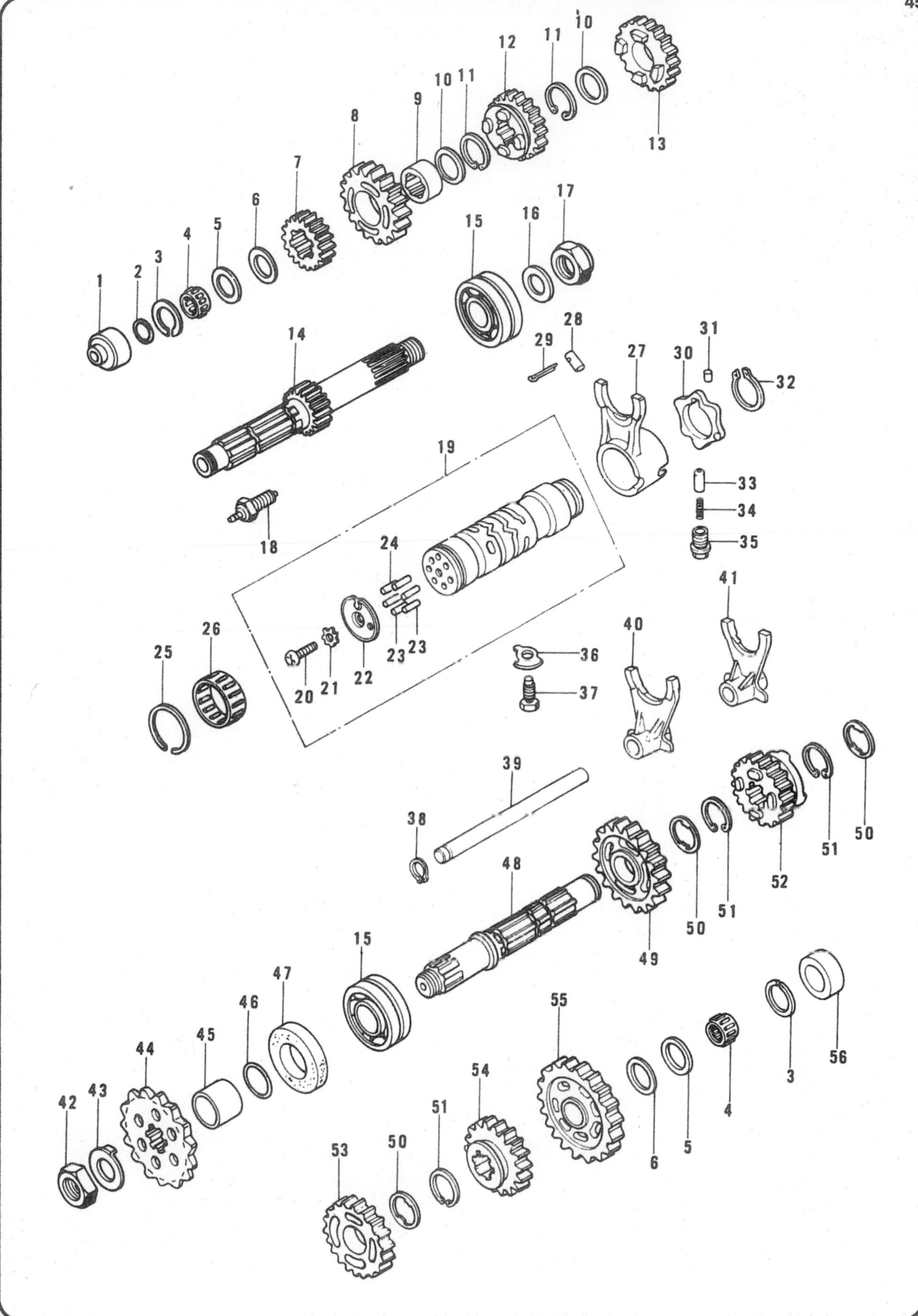
1
2
3
4
5
6
7
8
9
10
11
12
11
10
13
14
15
16
17
18
19
20
21
22
23
23
24
25
26
27
28
29
30
31
32
33
34
35
36
37
38
39
40
41
42
43
44
45
46
47
15
48
49
50
51
52
51
50
53
50
51
54
55
6
5
4
3
56

30 Engine reassembly: general

1 Before reassembly of the engine/gear unit is commenced, the various component parts should be cleaned thoroughly and placed on a sheet of clean paper, close to the working area.
2 Make sure all traces of old gaskets have been removed and that the mating surfaces are clean and undamaged. One of the best ways to remove old gasket cement is to apply a rag soaked in methylated spirit. This acts as a solvent and will ensure that the cement is removed without resort to scraping and the consequent risk of damage.
3 Gather together all the necessary tools and have available an oil can filled with clean engine oil. Make sure all new gaskets and oil seals are to hand, also all replacement parts required. Nothing is more frustrating than having to stop in the middle of a reassembly sequence because a vital gasket or replacement has been overlooked.
4 Make sure that the reassembly area is clean and that there is adequate working space. Refer to the torque and clearance settings wherever they are given. Many of the smaller bolts are easily sheared if over-tightened. Always use the correct size screwdriver bit for the crosshead screws and never an ordinary screwdriver or punch. If the existing screws show evidence of maltreatment in the past, it is advisable to renew them as a complete set.

31 Engine and gearbox reassembly: replacing the crankshaft

1 Check that all the bearing shells are laid out in the correct order, then refit them to their respective recesses. Ensure that the locating tang on each shell corresponds with the depression in which it engages. Ensure that each shell is firmly located before proceeding further. Oil each shell liberally.
2 Invert the upper crankcase and place it on the workbench. Commence reassembly by lowering the crankshaft assembly into position, taking care to locate all the holes in the outer tracks of the main bearings with the corresponding dowels located in the crankcase. Make sure the cam drive chain and Morse primary chain are in position on the crankshaft at this stage. Feed the connecting rods through the apertures in the crankcase and once the crankshaft is in position, rotate it to make sure all the main bearings revolve freely.
3 If the crankshaft does not appear to seat correctly, check that the crankshaft seals, which locate in grooves in the crankcase bosses, arل seating properly.

31.1a Fit the main bearing shells into position, noting locating tag

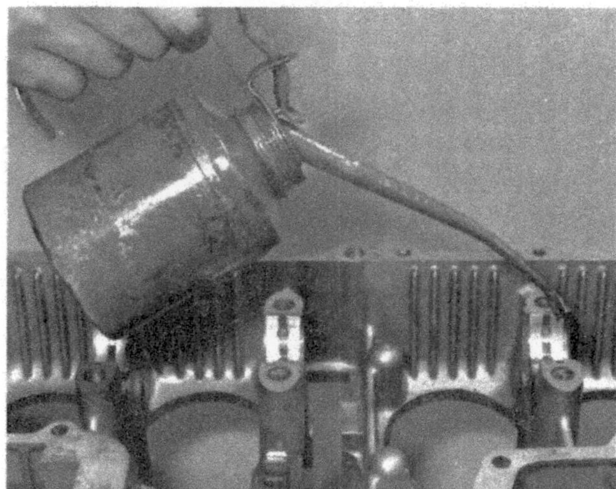
31.1b Lubricate bearing shells before fitting crankshaft

31.2a Fit new crankshaft oil seals ...

31.2b ... ensuring that they locate correctly in grooves when fitted

32 Reassembling the lower crankcase half components

1 Refit the kickstart mechanism from the inside of the case, sliding the sleeve into the casing and over the shaft, and retaining it with its locking plate and circlip.
2 The selector drum and third gear fork should be fitted next, following the removal sequence in reverse (See Section 17). Do not omit the detent plunger and locating bolt. Fit the selector forks and shaft, ensuring that they are in their correct positions.
3 Ensure that the main oil gallery is cleaned out. This can be achieved by removing the large plug at each end and blowing it out with compressed air. Refit and tighten the end plugs.

32.1 Refit kickstart mechanism, staking retaining screws

32.2a Refit selector drum bearing, if necessary ...

32.2b ... then refit selector drum, camplate and circlip

32.2c Fit pins in selector fork, and retain with split pin

32.2d Check that detent plunger assembly functions correctly ...

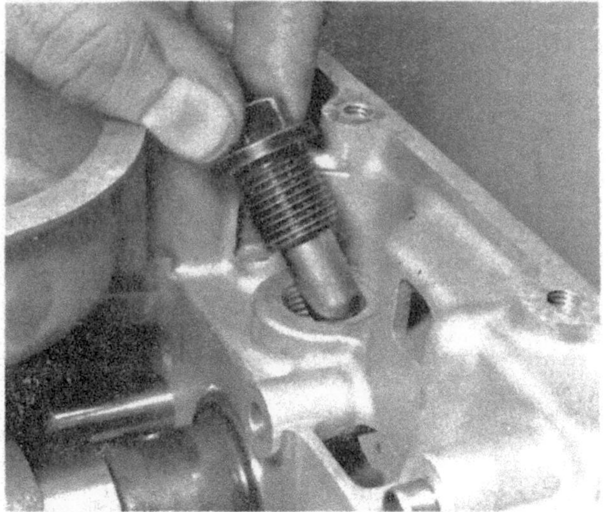

32.2e ... and refit in crankcase as shown

32.2f Refit and tighten locating bolt, and bend over locking tab

32.2g Check that mechanism operates smoothly, select neutral

32.2h Do not omit this circlip on selector fork shaft

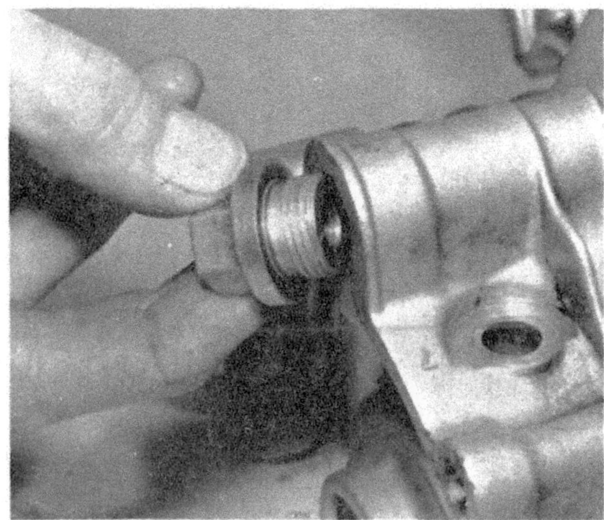

32.3a Remove the main oil gallery end plugs ...

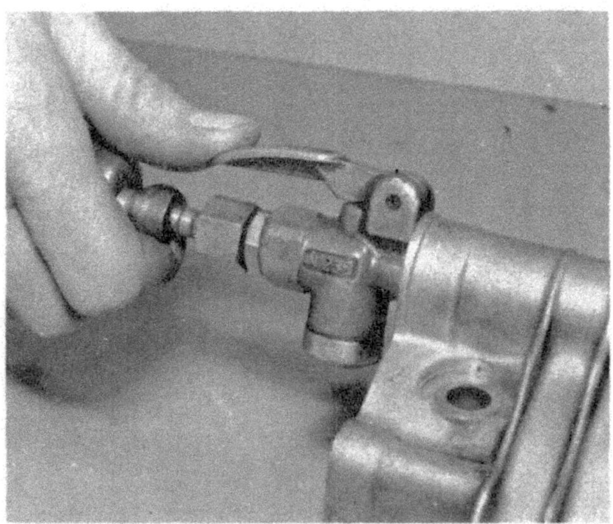

32.3b ... and clean out using compressed air

33 Engine and gearbox reassembly: refitting the gearbox clusters

1 The gearbox mainshaft and layshaft assemblies should be placed in position in the upper crankcase half. Note that each shaft is located by a half ring at one end, and a bearing locating pin at the other. Ensure that these engage correctly, and that the shafts seat squarely. Lubricate the assembly with clean engine oil, and check that the gears turn and mesh correctly.
2 Check, particularly if any gearbox components have been renewed, that the mainshaft 1st gear and layshaft 2nd gear pinions turn freely. If this is not the case, it will be necessary to fit a thinner thrust washer between the pinion concerned and its adjacent bearing. It is helpful if neutral is found by moving the gears so that drive between the mainshaft and layshaft is disconnected. Set the selector drum in the lower casing half in the neutral position. When set correctly, the smallest indentation in the cam plate will engage with the detent plunger.

34 Engine and gearbox reassembly: joining the upper and lower crankcase halves

1 Wipe off the mating surfaces of the crankcase halves to remove any residual oil. Check, and if necessary refit, any of the locating dowels which may have come loose.
2 Apply a thin film of non-hardening gasket compound to the mating surface of the lower crankcase half, taking care not to obstruct the oil passages. When an oilway is encountered, leave a narrow margin around it, so that the gasket compound will not get squeezed into it when the joint is tightened. If the sealing face has been marred or scratched, and oil leakage has been a problem in the past, it is permissible to use one of the Silicon RTV (Room Temperature Vulcanising) liquid gasket compounds now available. The rubbery nature of this substance will take up any small discrepancy in the mating face, but additional care must be taken to avoid obstructing oilways. Place new crankshaft seals in position, ensuring that they locate correctly.
3 Lower the lower crankcase half into position, checking to ensure that the selector forks engage in their respective grooves. Push the casing down evenly, ensuring that the locating dowels align correctly. Fit the ten 8 mm and eight 6 mm retaining bolts in place, finger-tight only.
4 The correct tightening sequence is stamped on the lower crankcase half, each of the 8 mm bolts having a number denoting in which order they are to be tightened. Set a torque wrench to about 1.5 kg m (11 ft lb) initially, and tighten the 8 mm bolts in sequence. Next reset the torque wrench to 2.5 - 3.0 kg m (18 - 22 ft lb) and re-tighten the bolts to their final torque setting. The remaining eight 6 mm bolts should be tightened evenly and in a diagonal sequence to 0.9 - 1.1 kg m (78 - 95 in lb) of torque.
5 Check at this stage that the crankshaft will turn freely and that there are no tight spots in evidence. The gearbox assembly should also be checked carefully for signs of resistance or rubbing, and the gear selection checked by turning the camplate to select each ratio. This checking sequence is important, as it is far preferable to spot any problems at this stage, rather than find them when reassembly is complete.
6 It is likely that the crankshaft will be a little stiff in its movement, particularly if new shells have been fitted. However, any undue resistance in either the crankshaft or gear components will necessitate rectification before rebuilding is continued. Turn the unit over and fit the twelve upper crankcase bolts, tightening them to 0.9 - 1.1 kg m (78 - 95 in lb).

35 Engine and gearbox reassembly: refitting the secondary shaft, oil pump and sump

1 Lower the secondary shaft sprocket/starter clutch assembly into the casing, and fit the Morse chain around the sprocket. Slide the secondary shaft into position, guiding the end of the shaft through the sprocket/clutch unit boss. Before driving the bearing home, fit the retainer plate, and hold this in position while the shaft is tapped into position. Refit the countersunk retaining screws, and remember to peen these into position after tightening.
2 Place the oil pump in position in the casing, then fit and tighten the mounting bolt and the two long screws which also retain the secondary shaft retaining plate. Note that the latter screws must be peened over after tightening.
3 Refit the secondary shaft bearing cap and tighten the mounting screws. Check that three new oilway 'O'-rings are fitted in their respective positions. Two of the 'O'-rings fit into recesses in the casing and a third is fitted in the base of the oil pump. Make sure that the lower casing and sump mating faces are clean, and that a new 'O'-ring is fitted to the oil filter chamber groove. A new sump gasket should be placed in position on the lower crankcase face, and the sump fitted and secured. Tighten the retaining bolts evenly to 0.7 - 0.9 kg m (61 - 78 in lb) of torque. A new oil filter can be fitted at this stage, and the cover fitted and tightened. It is worth priming the system by filling the inverted filter chamber with engine oil. This will speed up oil circulation when the engine is first started.

33.1a Fit half-rings and dowels in bearing journals

33.1b Refit gearbox clusters, and check operation

34.1 Crankshaft assembly may now be lowered into position

35.1a Guide primary chain over sprocket teeth

35.1b Slide secondary shaft partly home

35.1c Refit retainer, tightening and peening screws

35.2a Lower oil pump into casing ...

35.2b ... and fit bolt and screws. Peen the latter to prevent loosening

35.3a Bearing cap may now be refitted. Note O ring

35.3b Fit new O rings to oil pump and rear of casing ...

35.3c ... and also to the front of the casing

35.3d Oil filter chamber is sealed by large O ring

35.3e Place a new sump gasket in position and refit sump

36 Engine and gearbox reassembly: refitting the clutch assembly and contact breaker unit

1 Fit the special washer onto the end of the gearbox mainshaft, noting that the chamfered face must be innermost. The clutch needle roller bearing can be fitted next, followed by the clutch drum and thrust washer. Slide the clutch centre into position, followed by the special washer. Note that the latter is marked 'OUTSIDE' and must be fitted accordingly. Fit and tighten the clutch centre nut to 12-15 kg m (87-108 ft lb), locking the crankshaft in a similar manner to that used to hold it during dismantling.

2 Place the clutch plates in position, starting and finishing with a friction plate. Before fitting the cover, grease and insert the $\frac{3}{8}$ in steel ball in the centre of the mainshaft, followed by the mushroom-headed pushrod. The cover can now be refitted, the springs and washers placed in position, and the retaining bolts tightened down diagonally to 0.9-1.1 kg m (78-95 in lb) of torque.

3 If the kickstart shaft oil seal or (less likely) the oil level window, has shown signs of leaking, they should be driven out of the clutch cover. New seals can be carefully tapped back into position, taking great care not to damage the sealing edge.

4 The cover should be fitted using a new gasket. To prevent oil leakage, a smear of gasket cement should be applied at the two points where the crankcase halves are joined. Offer up the outer casing and tighten down the securing screws firmly.
5 Place the automatic timing unit in position on the end of the crankshaft, ensuring that the driving pin engages in the corresponding hole at the back of the unit. Fit and tighten the retaining bolt to 2.3-2.7 kg m (16.5-19.5 ft lb), holding the crankshaft with a 17mm spanner.
6 Connect the oil pressure switch lead (blue and red) to the terminal on the top of the sender. It is important that the tag is positioned well away from the contact breaker backplate, as it can easily short out the switch, giving a spurious indication of an oil pressure failure. Fit the contact breaker assembly in position, but do not fit the cover as the ignition timing must be checked after installation but before the cylinder head cover has been fitted. (For further details, refer to the Routine Maintenance section and Chapter 3).

37 Engine and gearbox reassembly: refitting the selector mechanism and alternator assembly

1 Check that the circlip is correctly located in its groove in the end of the selector fork shaft. Offer up the selector claw assembly, guiding the pawls each side of the selector drum end. Make sure that the centring spring engages on the locating pin. It is worthwhile temporarily refitting the gear change pedal to check that gear selection is positive.
2 Examine the seals in the outer cover. These can be renewed as necessary by driving the old seal out using a suitably sized socket as a drift. The casing should be arranged to give as much support as possible to the boss. The new seals may be driven into place in the same manner, taking care not to damage the sealing lip.
3 Slide a new 'O' ring into position on the end of the layshaft, and place a new gasket in position. Offer up the casing, easing the seals into position. Fit and tighten the retaining screws, not forgetting to position the sprocket guard. The sprocket spacer can now be slid into position, as can the clutch pushrod. Lubricate the two latter components with high melting point grease, prior to fitting.
4 Place the alternator rotor in position on the crankshaft taper, then fit and tighten the securing bolt to 5.8-6.3 kg m (42-46 ft lb). The crankshaft may be prevented from turning by way of the hexagon on the opposite end of the crankshaft, or by the method described in the removal sequence.
5 Ensure that the dowels and gasket are correctly positioned, then fit the outer cover, Do not forget to reconnect the neutral indicator switch lead.

36.1a Note that the chamfered face must fit against bearing

36.1b Fit chamfered washer and inner bearing race ...

36.1c ... followed by caged needle roller bearing, ...

36.1d ... clutch outer drum and thrust washer

36.1e Slide the clutch centre into position

36.1f Special washer must be fitted as shown

36.1g Lock clutch while centre nut is tightened

36.2a Refit the clutch plain and friction plates

36.2b Do not forget to grease and refit ball and pushrod

36.2c Fit the clutch cover, and replace the springs

36.2d Tighten the clutch retaining bolts

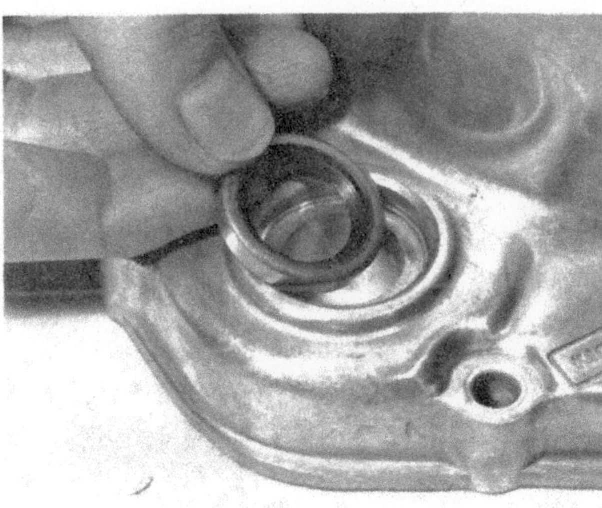
36.3 Renew any seals which appear worn or damaged

36.5 Note roll pin which engages in slot in base of ATU

37.1 Reassemble gear selector mechanism as shown

37.2a Fit new seals in cover as required ...

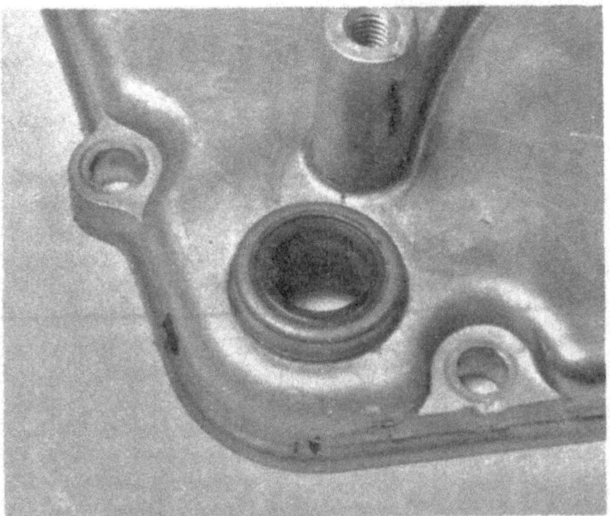
37.2b ... to avoid any risk of subsequent oil leakage

38 Engine and gearbox reassembly: replacing the pistons and cylinder block

1 Before replacing the pistons, pad the mouths of the crankcase with rag in order to prevent any displaced component from accidentally dropping into the crankcase.
2 Fit the pistons in their original order with the arrow on the piston crown pointing toward the front of the engine.
3 If the gudgeon pins are a tight fit, first warm the pistons to expand the metal. Oil the gudgeon pins and small end bearing surfaces, also the piston bosses, before fitting the pistons.
4 Always use new circlips, **never** the originals. Always check that the circlips are located properly in their grooves in the piston boss. A displaced circlip will cause severe damage to the cylinder bore, and possibly an engine seizure.
5 Place a new cylinder base gasket (dry) over the crankcase mouth. Refit the bottom guide roller. Now place the cylinder block over the cylinder studs (make sure the four 'O' rings are fitted to the base of the cylinders), support the cylinder block whilst the camshaft chain is threaded through the tunnel between the bores. This task is best achieved by using a piece of stiff wire to hook the chain through, and pull up through the tunnel. The chain must engage with the crankshaft drive sprocket.
6 The cylinder bores have a generous lead in for the pistons at the bottom, and although it is an advantage on an engine such as this to use the special Kawasaki ring compressor, in the absence of this, it is possible to gently lead the pistons into the bores, working across from one side. Great care has to be taken NOT to put too much pressure on the fitted piston rings. When the pistons have finally engaged, remove the rag padding from the crankcase mouths and lower the cylinder block still further until it seats firmly on the base gasket.
7 Take care to anchor the camshaft chain throughout this operation to save the chain dropping down into the crankcase. The two idler assembly sprockets that guide the cam chain, can now be replaced, with their shafts and rubbers, into the top of the cylinder block.

39 Engine and gearbox reassembly: replacing the cylinder head and camshafts

1 Rotate the crankshaft until the 'T' mark on the advance and retard mechanism is aligned with the timing mark as shown in the accompanying photograph. At this position, number one and four pistons are at top dead centre.
2 Replace the valve tappets and the shims in their original locations and use a new cylinder head gasket to prevent any compression leakage. Note that if the cylinder head, camshaft or tappets have been renewed, the tappet shims will almost certainly require changing. The valve clearances must, of course be checked in any event. The cylinder head can now be bolted down, tightening the nuts diagonally.
3 After the cylinder head has been secured, the next operation is to fit the camshafts. Start by fitting the exhaust camshaft first. To fit the camshaft, feed the camshaft through the cam chain, and turn the camshaft so that the mark on the sprocket is aligned with the cylinder head surface, as shown in the accompanying photograph.
4 Now pull the cam chain taut and fit the chain on to the camshaft sprocket. Starting with the next chain link pin above the one that coincides with the sprocket mark, count the pins, until you reach the 36th pin and slide the inlet camshaft into position so that the 36th pin coincides with the 'T' mark on the rubber portion of the inlet camshaft sprocket.
5 Having assembled the camshafts and replaced the cam chain, the next task is to bolt the camshafts down.
6 Lubricate the camshafts thoroughly, then fit the bearing caps in position. The caps are machined 'in line' with the cylinder head, and so it is very important that they are replaced with the number on the cap corresponding to the number on the cylinder head. Also the arrows marked on the caps must point to the front.
7 Partially tighten the left-hand caps first, to seat the camshafts in place. All the bolts can now be fully tightened down to 1.1-1.3 kg-m (95-113 in lb) of torque. They should be tightened down in numerical sequence.
8 Make sure all the camshaft bearings and valve tappets are lubricated with clean engine oil. The top chain guide sprocket can now be installed. Adjust the cam chain tension by refitting the tensioner. When this is installed loosen the locknut on the adjuster so that the plunger rod is free to move, rotate the engine slowly a couple of times to make sure the spring loaded tensioner takes up the slack in the chain evenly, and then tighten the bolt first and reset the locknut. It will adjust to the correct tension automatically.
9 To make doubly sure that the timing is right, rotate the engine until pistons number one and number four are at TDC and check that both the mark on the exhaust camshaft sprocket and the mark on the inlet camshaft sprocket are aligned level with the cylinder head surface. This will indicate that the cam timing is correct. **CAUTION.** Always use a spanner on the large nut on the crankshaft when turning the engine over for timing purposes. **DO NOT** turn the engine by turning the camshaft sprockets.
10 Note that the valve timing check is critical, as any error in this setting, however small, can result in the valves hitting the pistons or each other, causing extensive (and expensive) engine damage. It is, therefore, advisable to be absolutely certain that the timing is set correctly at this stage. On no account rush this operation.
11 Check the valve clearances, following the procedure detailed in the Routine Maintenance section. If the clearances are outside those specified, it will be necessary to remove the camshafts and tappets in order to fit smaller or larger shims. Again, do not be tempted to skip this operation, as excessive clearance will result in noisy operation and impaired efficiency. Conversely, too small a clearance will rapidly burn out the valve concerned. Once set, the clearance will not need adjustment for long periods.

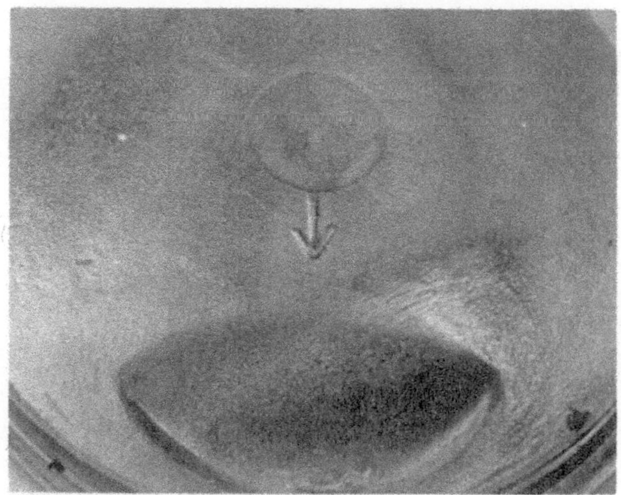

38.2 Arrow marking on pistons must face forward

38.4 Replace pistons, using new circlips

38.5a Fit a new cylinder base gasket and O rings

38.5b Make sure that dowels are positioned correctly

38.6 Feed pistons into cylinder bores, then lower block

39.1 Align T mark to set pistons 1 and 4 at TDC

39.2a Note that cylinder head gasket is marked TOP. Fit O rings

39.2b Lower cylinder head, supporting camshaft chain

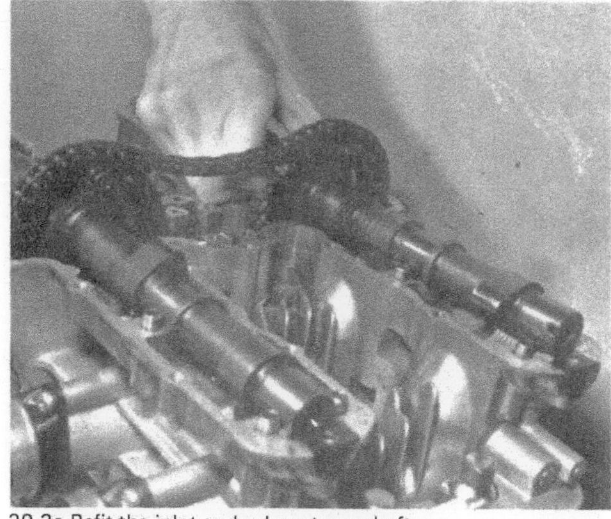
39.3a Refit the inlet and exhaust camshaft, ...

39.3b ... ensuring that the timing mark on each sprocket ...

39.3c ... is aligned as described in text

39.3d Crankshaft timing mark must be as shown when timing camshafts

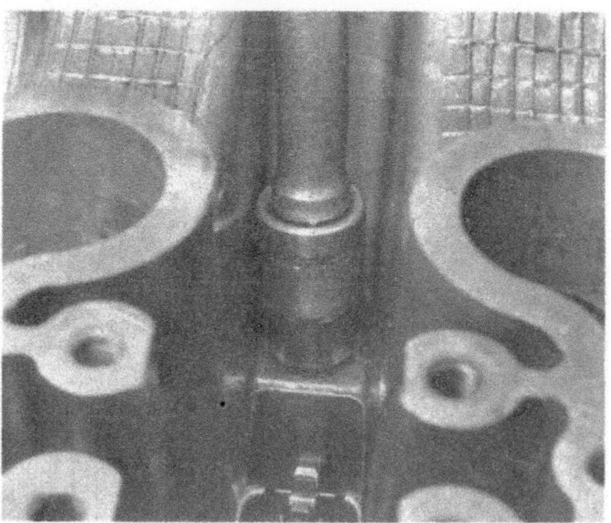
39.3e Do not omit recessed cylinder head bolt

39.6 Camshaft caps are numbered and arrowed for reference

39.8a Reassemble the upper chain guide sprocket assembly

39.8b Refit the camshaft chain tensioner, and adjust

39.8c Insert the tachometer drive and refit retaining plate

39.11a Check valve clearances, and reset where necessary

39.11b Shim sizes are etched on, see size chart

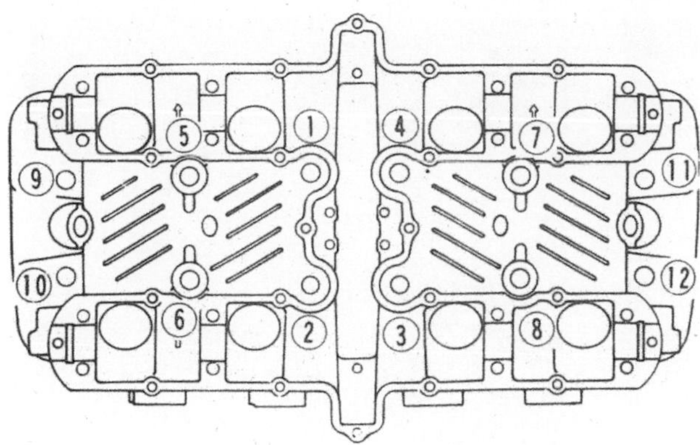

Fig. 1.10 Cylinder head nut tightening sequence

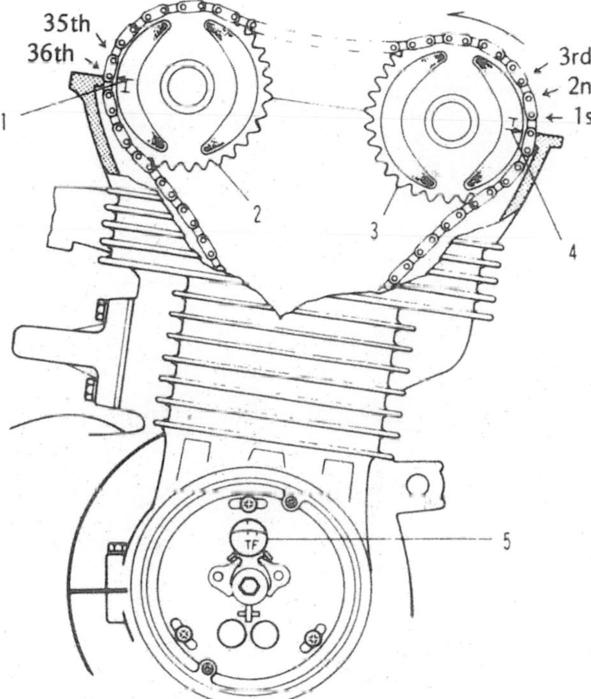

Fig. 1.11 Valve timing

1 Inlet camshaft timing mark
2 Inlet camshaft sprocket
3 Exhaust camshaft sprocket
4 Exhaust camshaft timing mark
5 Crankshaft timing mark (1 – 4 cylinder)

40 Refitting the engine and gearbox unit into the frame

1 As mentioned during the engine removal sequence, the engine/gearbox unit is unwieldy, requiring at least two, or preferably three, people to coax it back into position. This is even more important during reassembly, as the unit must be offered up at the right angle, and then manoeuvred into position. Care must be taken not to damage the finish on the frame tubes, and it is worthwhile protecting these with rag or masking tape.

2 Once the unit is sat on the frame cradle, refit the engine mounting bolts, referring to the photograph showing their location, which accompanies the removal sequence in the early part of this Chapter. Assemble the front and rear engine plates loosely and fit the nuts and spring washers finger tight. Lift the rear of the unit so that the upper rear bolt hole (3) is aligned, and slide the bolt into place, noting that a spacer is also required. Next, fit the front mounting bolt (1) and the two lower mounting bolts (2 and 4).

4 For ease of identification, note that the mounting bolt lengths are as follows:

Front upper mounting bolt:	296 mm (11·7 in) (No 1)
Front lower mounting bolt:	323 mm (12·7 in) (No 2)
Rear upper mounting bolt:	250 mm (9·8 in) (No 3)
Rear lower mounting bolt:	225 mm (8·9 in) (No 4)

The four engine plate bolts should be tightened to 2.0-2.8 kg m (14.5-20 ft lb) and the engine mounting bolts to 3.4-4.6 kg m (25-33 ft lb) torque.

41 Engine and gearbox unit installation: final assembly and adjustment

1 Place the final drive chain over the rear sprocket, and fit it over the splined end of the layshaft. Refit the tabwasher and nut, tightening the latter to 7.5-8.5 kg m (54-61 ft lb). Lock the rear wheel by applying the rear brake whilst the nut is tightened. Knock over the locking washer to prevent the nut from slackening in use.

2 Reconnect the clutch cable, and refit it to the frame using cable ties. Check that the neutral switch lead has been reconnected. Fit the starter motor lead to the terminal, and slide the protective rubber boot into position to protect the terminal from shorting. Offer up the starter motor, taking care not to damage the 'O' ring, and refit the two mounting bolts.

3 Check that the alternator output leads are routed correctly, then fit the outer casing and tighten the retaining screws. Refit the left-hand footrest, then fit the gearchange pedal, checking that it is in the right position in relation to the footrest. Assemble the right-hand footrest, reconnect the brake light switch operating spring and leads. If necessary, readjust the rear chain tension, brake pedal free play and rear brake switch operation.

4 Check the ignition timing as described in the Routine Maintenance section or Chapter 3, then refit the cylinder head cover and contact breaker cover. Check that the contact breaker leads (blue/red, black and green) and the alternator output leads (multiple connector and green lead) are reconnected and correctly routed.

5 Make sure the holding plate is secured tightly to the four carburettors with the eight countersunk screws before replacing the whole carburettor bank. Also check that the throttle control cable wheel operates and returns freely on the return spring, and that the choke lever operates the chokes of all four carburettors. Fit the carburettor bank to its inlet stubs, with the aid of an assistant, if possible.

6 Secure the carburettors to the intake hoses on the cylinder head by the securing clips fitted round the intake hoses. Make sure these clips are tight, otherwise leakage will occur on the intake side of the carburation and cause irregular running.

7 Channel the four rubber overflow pipes through the retaining band at the rear of the engine, adjacent to the oil filler cap.

8 Refit the air cleaner housing mounting brackets, and secure them with the single mounting bolts at each side. Replace the breather cover cap if this has been detached, then reconnect the large breather pipe, routing it behind the crankcase to the air cleaner housing.

9 Reassemble the ignition coils and refit the numbered sparking plug leads to their appropriate plugs. Refit each half of the exhaust system, using new exhaust port gaskets. Note that the half-clamps can be held in place with strips of masking tape while the clamps are refitted. Make sure that the balance pipe which runs beneath the crankcase is pushed together firmly, and the clamp bolts tightened securely. Refit the petrol tank and reconnect the petrol feed pipe to the carburettor. Reconnect the tachometer drive cable.

10 Check that the clutch cable is correctly adjusted, (See Routine Maintenance) and that the throttle cables operate smoothly and without excessive play (See Chapter 2). Reconnect the battery, noting that the system is negative (-) earthed, then refit the side panels. Refill the crankcase with SAE 10W/40 or 10W/50 engine oil, bringing the level to halfway up the oil level window. Note that this level should be checked after running the engine for a few minutes, and replenished as necessary. Check around the machine for any components which may have been overlooked, and check the workbench for 'spare' parts - there should be none left over.

40.2a Assemble the engine front plates ...

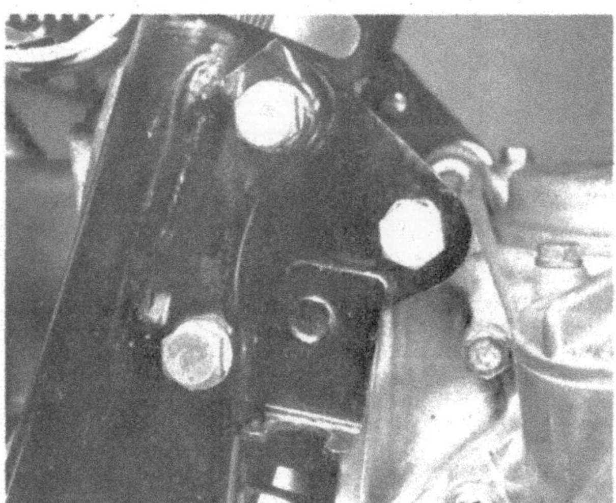
40.2b ... and the engine rear plates, do not tighten yet

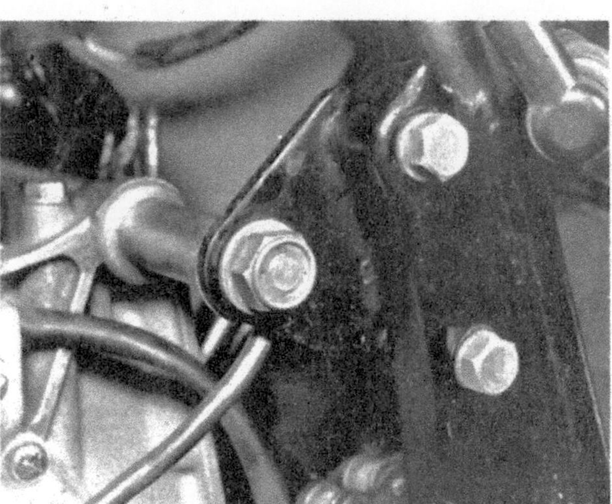
40.2c Note spacer between rear left-hand plate and engine

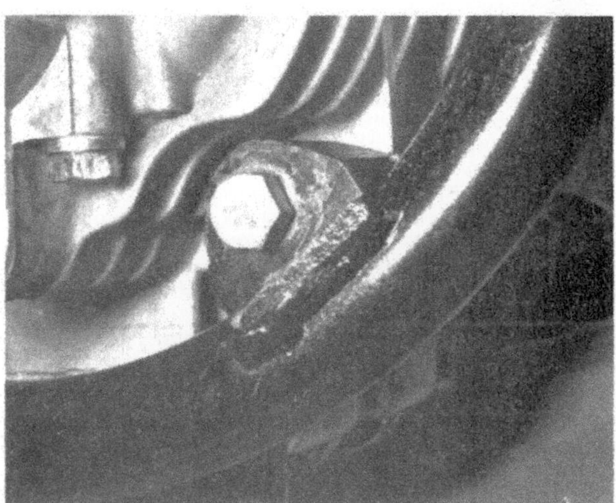
40.2d Refit the front lower mounting bolt ...

40.2e ... and rear lower mounting bolt

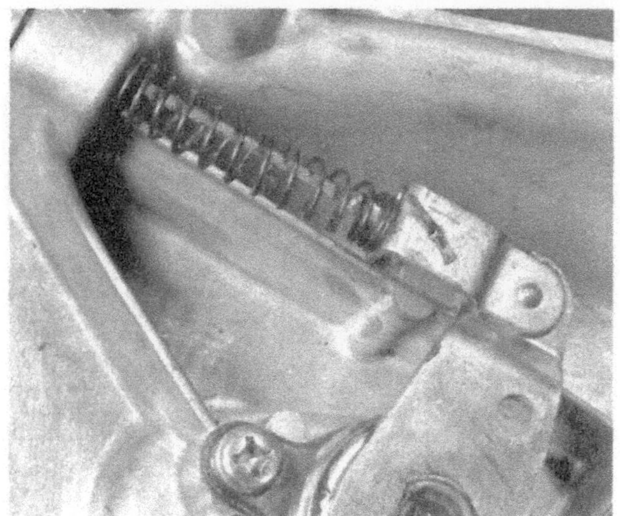
41.2a Reconnect the clutch cable before refitting clutch cover

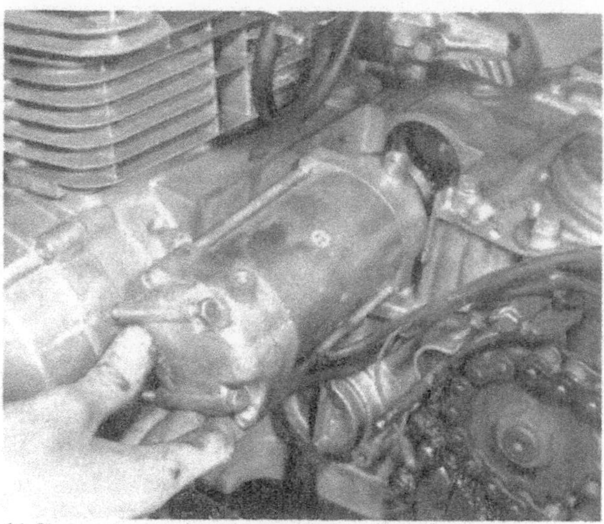
41.2b Attach starter motor lead before installing motor

41.3a Alternator cover and stator assembly can now be refitted

41.3b Do not forget to fit the engine earth lead

41.4 Refit the cylinder head cover after re-timing the ignition

41.8 Air cleaner trunking is retained by bolt at each side

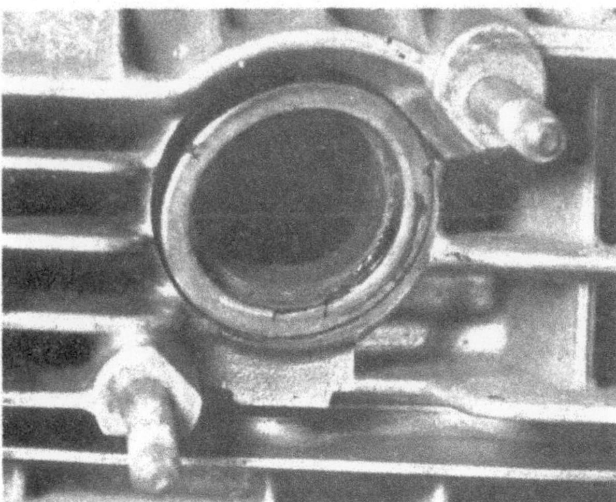

41.9a New exhaust port gaskets can be held in place with grease

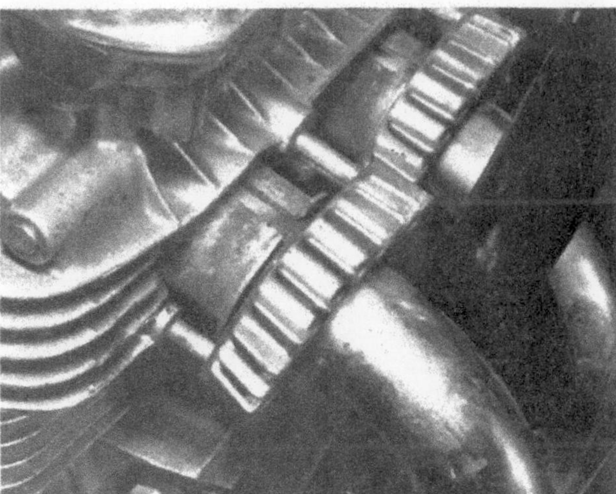

41.9b If necessary, hold split retainers by binding with tape

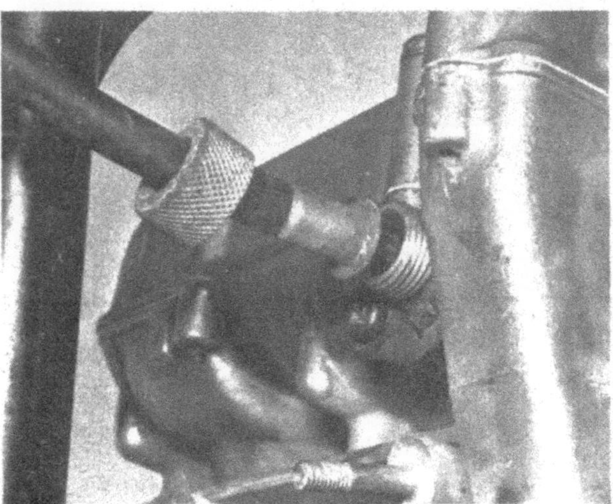

41.9c Reconnect the tachometer drive cable

41.10 Top up engine oil to mid point of level window

42 Starting and running the rebuilt engine unit

1 Make sure that all the components are connected correctly. The electrical connectors can only be fitted one way, as the wires are coloured individually. Make sure all the control cables are adjusted correctly. Check that the fuse is in the fuse holder. try all the light switches and turn on the ignition switch. Close the choke lever to start.

2 Switch on the ignition and start the engine by turning it over a few times with the kickstart or the electric starter, bearing in mind that the fuel has to work through the four carburettors. Once the engine starts, run at a fairly brisk tick-over speed to enable the oil to work up to the camshafts and valves.

3 Before taking the machine on the road, check that the brakes are correctly adjusted, with the required level of hydraulic fluid in the handlebar master cylinder.

4 Make sure the rear chain is correctly tensioned to $\frac{3}{8}$ inch up and down play. Also that the front forks are filled with the correct amount of oil.

5 Check the exterior of the engine for signs of oil leaks or blowing gaskets. Before taking the machine on the road for the first time, check that all nuts and bolts are tight and nothing has been omitted during the reassembling sequence.

43 Taking the rebuilt machine on the road

1 Any rebuilt engine will take time to settle down, even if the parts have been replaced in their original order. For this reason it is highly advisable to treat the machine gently for the first few miles, so that the oil circulates properly and any new parts have a reasonable chance to bed down.

2 Even greater care is needed if the engine has been rebored or if a new crankshaft and main bearings have been fitted. In the case of a rebore the engine will have to be run-in again as if the machine were new. This means much more use of the gearbox and a restraining hand on the throttle until at least 500 miles have been covered. There is not much point in keeping to a set speed limit; the main consideration is to keep a light load on the engine and to gradually work up the performance until the 500 mile mark is reached. As a general guide, it is inadvisable to exceed 4,000 rpm during the first 500 miles and 5,000 rpm for the next 500 miles. These periods are the same as for a rebored engine or one fitted with a new crankshaft. Experience is the best guide since it is easy to tell when the engine is running freely.

3 If at any time the oil feed shows signs of failure, stop the engine immediately and investigate the cause. If the engine is run without oil, even for a short period, irreparable engine damage is inevitable.

44 Fault diagnosis: engine

Symptom	Cause	Remedy
Engine will not start	Defective sparking plugs	Remove the plugs and lay them on the cylinder head. Check whether spark occurs when engine is on and engine rotated.
	Dirty or closed contact breaker points	Check the condition of the points and whether the points gap is correct.
	Faulty or disconnected condenser	Check whether the points arc when separated. Renew the condenser if there is evidence of arcing.
Engine runs unevenly	Ignition or fuel system fault	Check each system independently, as though engine will not start
	Blowing cylinder head gasket	Leak should be evident from oil leakage where gas escapes.
	Incorrect ignition timing	Check accuracy and reset if necessary.
Lack of power	Fault in fuel system or incorrect ignition timing	Check fuel lines or float chambers for sediment. Reset ignition timing.
Heavy oil consumption	Cylinder block in need of rebore	Check bore wear, rebore and fit oversize pistons if required.

45 Fault diagnosis: clutch

Symptom	Cause	Remedy
Engine speed increases as shown by tachometer but machine does not respond	Clutch slip	Check clutch adjustment for free play, at handlebar lever, check thickness of inserted plates.
Difficulty in engaging gears, gear changes jerky and machine creeps forward when clutch is withdrawn difficulty in selecting neutral	Clutch drag	Check clutch for too much free-play. Check plates for burrs on tongues or drum for indentations. Dress with file if damage not too great.
Clutch operation stiff	Damaged, trapped or frayed control cable	Check cable and renew if necessary. Make sure cable is lubricated and has no sharp bends.

46 Fault diagnosis: gearbox

Symptom	Cause	Remedy
Difficulty in engaging gears	Selector forks bent	Replace with new forks.
	Gear clusters not assembled correctly	Check gear cluster for arrangement and position of thrust washers.
Machine jumps out of gear	Worn dogs on the ends of gear pinions	Renew worn pinions.
Gear change lever does not return to original position	Broken return spring	Renew spring.
Kickstarter does not return when engine is turned over or started	Broken or wrongly tensioned return spring	Renew spring or retension.
Kickstarter slips	Ratchet assembly worn	Dismantle engine and renew all worn parts.

Chapter 2 Fuel system and lubrication

Contents

Specifications

Petrol tank capacity

Total:	16.8 litres (3.7 Imp gall/4.4 US gals)
Reserve:	2.4 litres (0.5 Imp gall/0.6 US gals)

Carburettors

Make	Mikuni
Type	VM24SS
Main jet	100
Pilot jet	16
Needle jet	0-8
Jet needle	*5DL 31-4
Throttle valve cutaway	1.5
Pilot screw setting	$\frac{3}{4} \pm \frac{1}{8}$ turns out from stop.
Float level	30 $\pm$ 1mm nominal

***Note** 5DL indicates needle number, 31 is the manufacturers batch number, and -4 is the needle position, ie. 4 grooves from top

Oil capacity

Dry:	3.5 litres (6.2 Imp pints/3.7 US quarts)
Oil change:	3.0 litres (5.3 Imp pints/3.2 US quarts)

Oil pump

Type	twin rotor trochoidal
Inner rotor to outer rotor clearance:	0.05 - 0.23 mm (0.002 - 0.009 in) nominal 0.30 mm (0.012 in) wear limit
Outer rotor to pump body clearance:	0.15 - 0.21 mm (0.006 - 0.008 in) nominal 0.30 mm (0.012 in) wear limit
Rotor to end plate clearance:	0.02 - 0.07 mm (0.0008 - 0.0028 in) nominal 0.12 mm (0.005 in) wear limit
Oil pressure at 4,000 rpm/90°C (194°F):	2.0 - 2.5 kg/cm^2 (28 - 36 psi)

1 General description

The fuel system is comprised of a steel petrol tank, from which fuel is fed by gravity to the four Mikuni VM24SS carburettors. The tap has three positions, giving a normal supply of petrol, an emergency reserve position and an off position. A gauze strainer is incorporated in the tap, to trap any foreign matter which might otherwise block the carburettor jets. The petrol tank filler cap is of the quick-action type, incorporating a lock operated by the ignition key.

The four carburettors are interconnected by a linkage to ensure synchronisation. The two throttle cables are connected to a pulley mounted between the two control instruments, the throttles being opened and closed positively. The engine draws air in via a moulded plastic trunking which contains the air cleaner element.

The engine oil is contained in a sump formed at the bottom of the crankcase.

The gearbox is also lubricated from the same source, the whole engine unit being pressure fed by a mechanical oil pump that is driven off the crankshaft. The oil pump intake extends into the sump to pump the oil up to the engine. A screen at the pump inlet point prevents foreign matter from entering the pump before it can damage the mechanism. From the pump the oil passes to the oil filter to be cleaned. If the filter becomes clogged, a safety by-pass valve routes the oil around the filter. It is then routed through a passageway in which an oil pressure switch is mounted, and through an oil hole in the crankcase, from which point it is sent in three different directions. One direction is to the crankshaft main bearings and crankshaft pins. After lubricating the crankshaft parts, the oil is thrown out by centrifugal force and the spray lands on the cylinder walls, the pistons and gudgeon pins, to lubricate those parts. The oil eventually drops down from all these points and accumulates in the bottom of the crankcase sump to be recirculated.

The second passageway for oil from the pump is through the oil passage at each end of the cylinder block and up into the cylinder head. After passing through holes into the camshaft bearings, the oil flows out over the cams and down around the valve tappets to lubricate these areas. The oil returns to the sump via the oil holes at the base of the tappets, and the cam chain tunnel in the centre of the cylinder head and cylinder block.

The third passageway for the oil to flow is to the gearbox bearings where it is pumped to the gearbox main bearing on the mainshaft and also to the bearing on the layshaft. After this the oil drops down back into the oil sump, to be recirculated through the engine.

2 Petrol tank; removal and replacement

1 The petrol tank fitted to the Z650 4-cylinder models is secured to the frame by means of a short channel that projects from the nose of the tank and engages with a rubber buffer surrounding a pin welded to the frame immediately behind the steering head. This arrangement is duplicated either side of the nose of the tank and the frame. The rear of the tank is secured by a rubber clip that locates round a lip welded on to the back of the tank. The tank also has two rubber buffers on which it rests at the rear. A petrol tap is fitted with a reserve pipe that is switched over, when the fuel level falls below that of the main feed pipe.

2 The petrol tank can be removed from the machine without draining the petrol, although the rubber fuel lines to the carburettors will have to be disconnected. The dualseat must be lifted up to release the rubber clip at the rear of the tank, then the tank raised at the rear and pulled upwards and backwards to pull off the front rubbers. When replacing the fuel tank, lift at the rear and push down onto the front rubber buffers, then secure the rubber clip at the rear and reconnect the fuel lines.

3 Petrol tap and filter: removal, dismantling and replacement

1 It is not necessary to drain the petrol tank if it is only half or under half full, as the tank can be laid on its side on a clean cloth or soft material (to protect the enamel), so that the petrol tap is uppermost. The petrol pipe should be removed before unscrewing the petrol tap. To remove the tap and filter first undo the two mounting bolts next to the tank, the tap body can then be detached complete with the filter. When the filter bowl is removed this will reveal the rubber 'O' ring gasket and the filter gauze. Remove the gauze and clean in petrol. When reassembling the tap, fit a new gasket between the body and the tank and a new rubber 'O' ring to the filter bowl if the old one is noticeably compressed. On no account overtighten the bowl, as it is made of soft metal and the threads will easily strip.

2 There is no necessity to remove either the tap or the petrol tank if only the filter bowl has to be detached for cleaning.

3.1a Tap is retained to petrol tank by two bolts

3.1b Bowl can be unscrewed and detached ...

3.1c ... to expose gauze filter insert

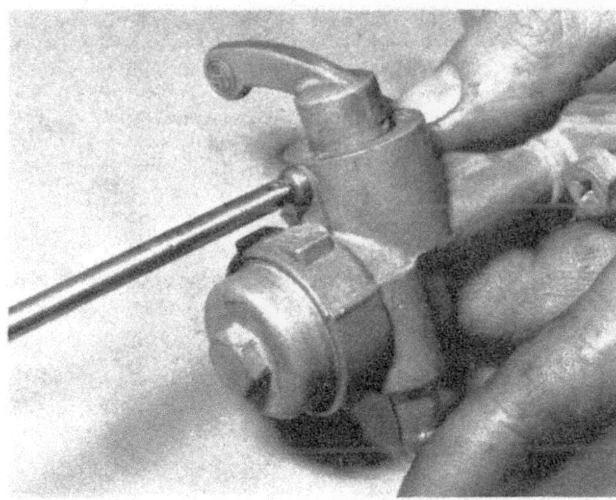
3.1d Single screw in tap body ...

3.1e ... retains the tap lever assembly

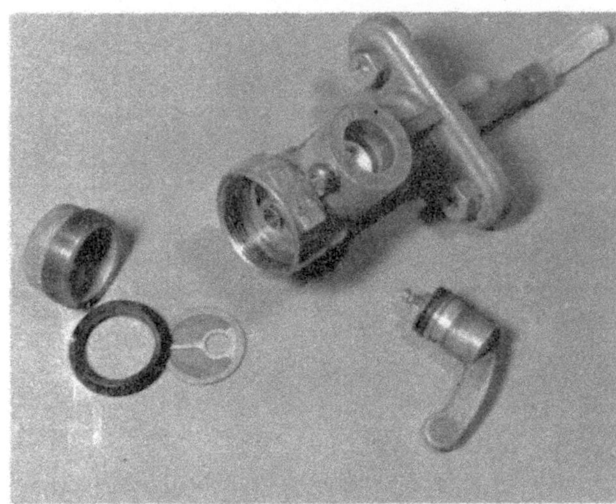
3.1f Examine and renew tap components as necessary

4 Carburettors: description and removal

1 The method of mounting the four carburettors is on a holding plate with eight countersunk screws, the whole bank of the carburettors being connected to the intake side of the engine on short induction hoses. The four plunger type throttle valves are operated by a single shaft, likewise the manually operated choke also has a single shaft operating four levers, one to each carburettor.

2 The throttles are operated by two cables from the handlebar, one to open the throttles and the other to close them. A heavy return spring is incorporated in the throttle return system. When closing the throttles the use of a separate return cable helps to close the throttle more positively. This ensures smooth throttle action.

3 A vacuum gauge fitting is incorporated in each inlet manifold as an aid to balancing manifold pressure, and when used in conjunction with a vacuum gauge array, allows the carburettors to be accurately synchronised.

4 Starting in extreme cold weather is aided by a separate starter system which acts by vacuum pressure and serves in place of a choke. The starter system takes the form of four plunger valves through the starter pipe to the starter plunger chamber, where fuel is atomized. The rich mixture is then sprayed into the carburettor bore where a small fuel/air spray from the pilot system is mixed with it. The final mixture is then delivered to the engine.

It is essential that the starter plungers be fully raised by the choke lever and the starter jet, pipe, and the air bleed hole completely free of any blockage. The throttle must be fully closed so that sufficient vacuum is developed for efficient atomization to take place.

6 The pilot system is made up of the pilot jet, the pilot air screw, and the pilot outlet. It controls carburation from the idle position to approximately one eighth throttle opening. The pilot mixture strength is determined by the amount of fuel passed through the pilot jet, and also by the amount of air which is allowed to pass the pilot air screw. If the screw is turned in, this richens the mixture, and when the screw is turned out this weakens the mixture. The correct position for the air screw is normally one and a quarter turns out from the fully closed position.

7 The main carburation system comprises the main jet, the bleed pipe, the jet needle, the throttle valve, and the air jet. The main system comes into operation after the throttle is opened

beyond one eighth of a turn. It is only after this that sufficient vacuum is created at the jet needle to draw fuel up through the main jet. The fuel flows up through the main jet and bleed pipe, then between the needle jet and jet needle, and into the main bore when it is fully atomized. The fuel in fact is partially atomized before it reaches the bore, because the air bleed hole in the bleed pipe admits air to the fuel as it passes through the pipe.

8 When the throttle is opened, the slide rises up the bore of the carburettor. The jet needle is connected to this slide and because the needle is tapered, the more it is raised the more the fuel is allowed to flow. This is how engine speed increases. The cutaway on the slide regulates the air flow and the vacuum pressure into the carburettor bore. Finally, when the throttle slide is raised to its limit, the flow of fuel is dictated by the size of the main jet rather than the space between the jet needle and needle jet. The machine is then running on the main jet.

9 The float system is made up of the float, the float needle valve, and the valve seat. The fuel is maintained by the float assembly at a constant level in the float bowl, to meet the engine's needs. As fuel flows into the bowl the float rises which in turn raises the float valve. When the float reaches a predetermined height, the valve closes onto its seat and this shuts off the flow of fuel to the carburettor. Consequently as the engine uses the fuel, the level in the bowl drops, causing the valve to leave its seat and admit more fuel to flow into the float chamber. Cleanliness is the most important thing when working on the carburettors. To remove the carburettors loosen the four intake manifolds by undoing the crosshead screws in the clamps and remove the air cleaner hoses at the rear. Then pull off the whole bank of four carburettors.

10 To separate the carburettors, loosen the throttle cable mounting nuts, and disconnect the cables from the pulley. Remove the throttle stop screw locknut from both the carburettor to be detached and its companion, then detach the link piece. Remove the throttle stop screw and the screw spring, together with the spring seat. Remove the cap nut from the carburettor linkage of the carburettor that is to be detached, and then remove the spring and seat. **Note**: Be careful not to lose the spring that will rise up when the cap nut is unscrewed.

11 Unscrew the four mounting screws from the mounting plate and remove the first pair of carburettors. It is easier to remove the carburettors in pairs as they are joined by a link. After all the carburettors have been removed from the mounting plate they are ready for dismantling.

12 Alternatively, the carburettors may be removed as an assembly as described in Chapter 1, Section 5, and then detached from their mounting plate.

4.12a Carburettor assembly can be removed as a unit

4.12b Remove carburettor top and gasket ...

4.12c ... to expose rocker assembly. Release clamp bolt

4.12d Each instrument is retained by two screws

4.12e Swing carburettor away from plate to disengage fork

4.12f Carburettor can be slid off coupling shaft

5 Carburettors: dismantling and reassembly

1 The crossover lever and pulley, and the throttle return spring need not be removed from the mounting plate, when dismantling the carburettors. The fuel may be drained from the float bowl by detaching the drain plug and washer. Remove the top cover screws, then remove the cover and gasket, bend flat the locktab washer and unscrew the bolt from the operating arm. The operating arm can now be removed. Undo the two screws that secure the bracket assembly to the throttle slide, and lift the bracket complete with the operating arm and connector assemblies out of the carburettor bore.

2 Remove the throttle valve and the needle from the bore taking care not to bend the needle. Remove the plunger assembly after first removing the lever, cap, and guide screw. Undo the float bowl screws, remove the bowl and the gasket, then take out the hinge pin and remove the float and the float needle valve. Remove the main jet, the air bleed pipe. Invert the carburettor, and gently press out the needle jet with a wooden dowel. Remove the float valve seat, the pilot jet and the pilot air screw with spring.

3 Clean all the components in clean petrol and then blow them dry with compressed air, taking care to clear all passages. Inspect all the jets and the needle valve and seat, and renew them if they are worn, especially if there is a bright ridge round the needle valve and seat. It is best to renew these as a pair.

4 Inspect the float for leakage. Check whether petrol has entered the float by shaking it. If the float assembly is punctured it must be renewed.

5 Remove the main jet with a wide blade screwdriver, also inspect the needle jet for wear. After lengthy service the needle jet should be renewed along with the needle as these components are in continuous use. If not renewed, petrol consumption will increase.

6 The carburettor slide should be able to slide down the carburettor bore by its own weight. If it will not do this, even when lightly oiled, it will not function correctly.

7 Assembling of the carburettors is the reverse order of dismantling. Use new gaskets and 'O' rings. Do not overtighten the jets when installing into the carburettor body.

8 Make certain that the carburettor jet needle is replaced back in the same position as when it was removed. The needle clip should be in the fourth groove from the top.

5.2a Lift out rocker and throttle valve assembly

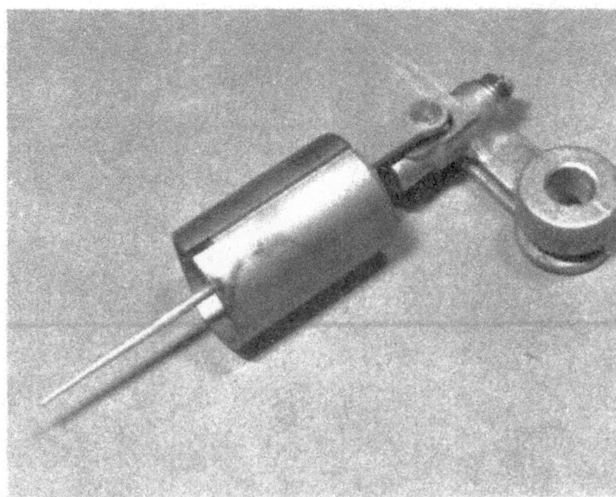
5.2b Examine valve assembly for wear or damage

5.2c Valve is retained by two screws to linkage

5.2d Linkage may be released when adjuster is unscrewed

5.2e Needle can be shaken out of valve for examination

5.2f Release carburettor float bowl and lift clear

5.2g Pivot pin should be displaced to release float assembly

5.2h Needle will drop clear of the valve seat

5.2i Valve seat screws into carburettor body

5.2j Unscrew main jet ...

5.2k ... and air bleed tube

5.2l Pilot jet should be removed from adjacent bore

5.2m Pilot mixture screw assembly is located here

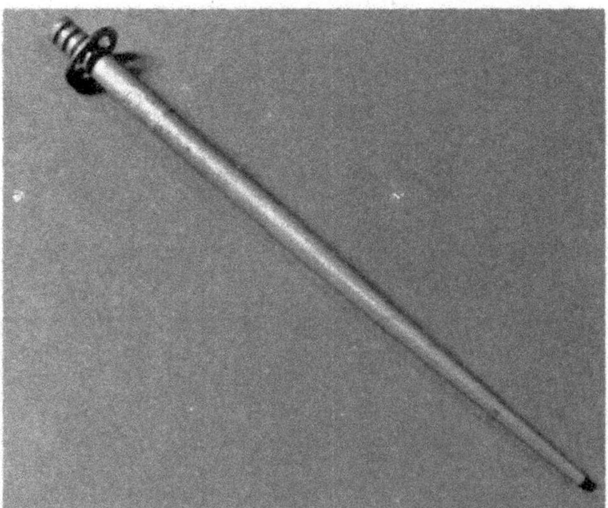
5.3a Check needle clip position. Renew if scored or bent

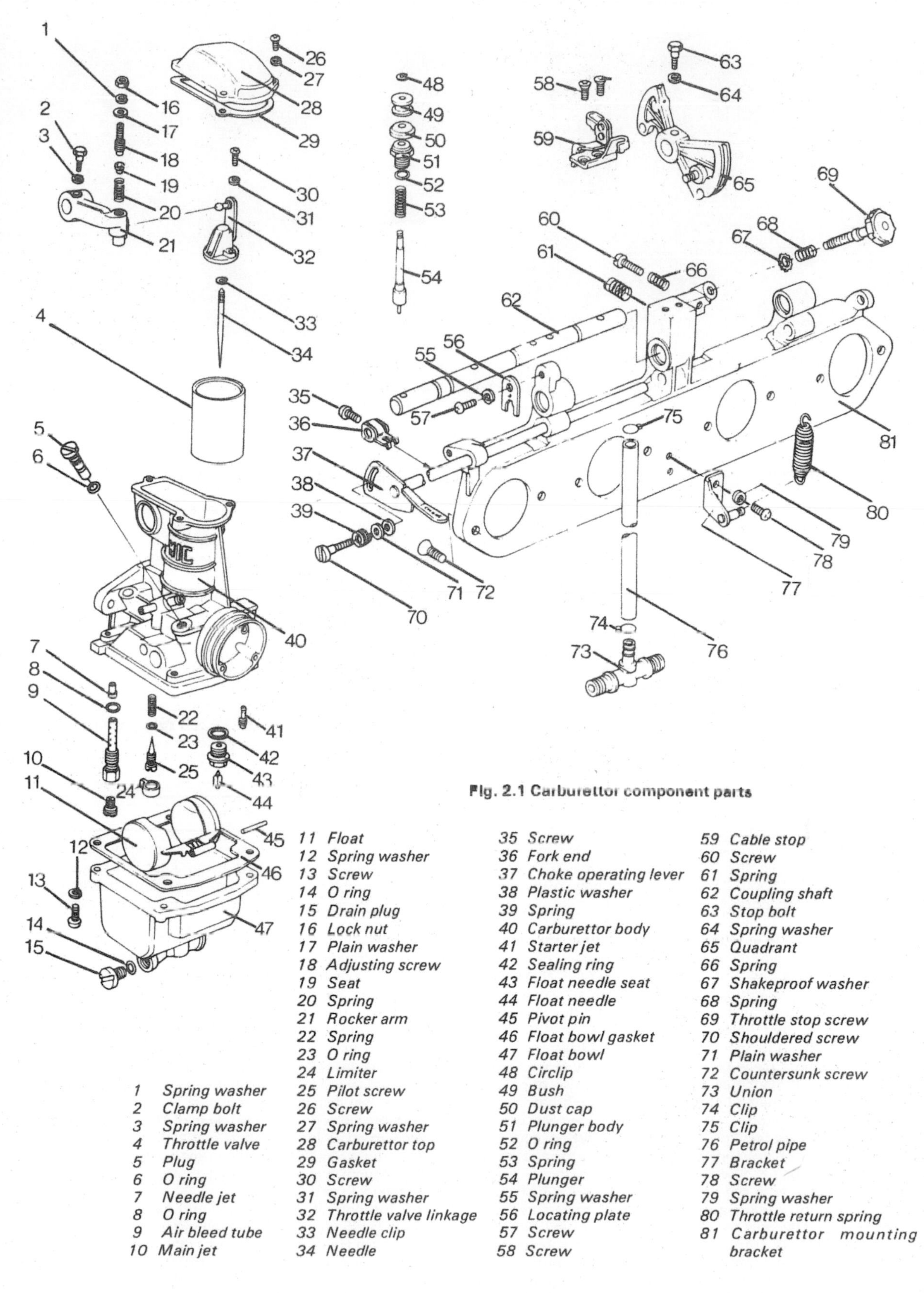

Fig. 2.1 Carburettor component parts

1 Spring washer
2 Clamp bolt
3 Spring washer
4 Throttle valve
5 Plug
6 O ring
7 Needle jet
8 O ring
9 Air bleed tube
10 Main jet
11 Float
12 Spring washer
13 Screw
14 O ring
15 Drain plug
16 Lock nut
17 Plain washer
18 Adjusting screw
19 Seat
20 Spring
21 Rocker arm
22 Spring
23 O ring
24 Limiter
25 Pilot screw
26 Screw
27 Spring washer
28 Carburettor top
29 Gasket
30 Screw
31 Spring washer
32 Throttle valve linkage
33 Needle clip
34 Needle
35 Screw
36 Fork end
37 Choke operating lever
38 Plastic washer
39 Spring
40 Carburettor body
41 Starter jet
42 Sealing ring
43 Float needle seat
44 Float needle
45 Pivot pin
46 Float bowl gasket
47 Float bowl
48 Circlip
49 Bush
50 Dust cap
51 Plunger body
52 O ring
53 Spring
54 Plunger
55 Spring washer
56 Locating plate
57 Screw
58 Screw
59 Cable stop
60 Screw
61 Spring
62 Coupling shaft
63 Stop bolt
64 Spring washer
65 Quadrant
66 Spring
67 Shakeproof washer
68 Spring
69 Throttle stop screw
70 Shouldered screw
71 Plain washer
72 Countersunk screw
73 Union
74 Clip
75 Clip
76 Petrol pipe
77 Bracket
78 Screw
79 Spring washer
80 Throttle return spring
81 Carburettor mounting bracket

5.3b Coupling pipe is sealed with O rings

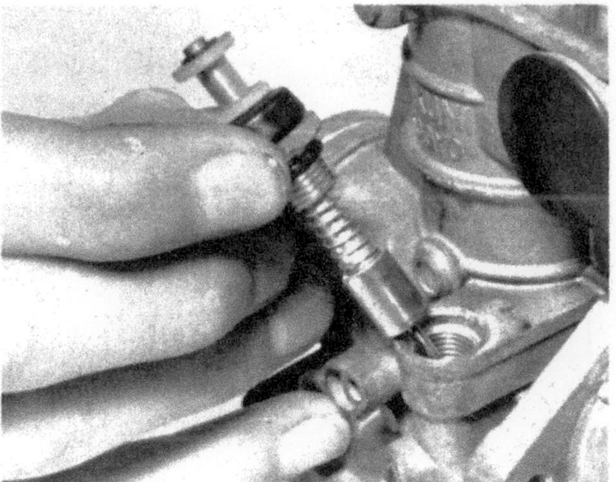

5.3c Enrichening device rarely requires attention

6 Carburettors: adjustment

1 To check the float height adjustment with the carburettors in situ, first turn the fuel tap to 'OFF'. Then, remove the carburettor vent tube. (Be prepared to catch the fuel that will run out). Remove the float bowl drain plug. Install the Kawasaki fuel measuring device (part number 57001-208) in place of the drain plug and hold the plastic tube against the carburettor body. Turn the fuel tap to the 'ON' position. The petrol level in the tube should be 2.5 - 4.5 mm (0.10 - 0.18 inch) below the edge of the carburettor body. If the petrol level is incorrect the float must be adjusted in the following manner. Drain the fuel from the float chamber and remove the chamber bowl. Be prepared to catch the float and float hinge pin, also the float needle. Bend the tang on the float slightly to adjust the float height. Bending the tang up will lower the fuel level. **Note:** When checking the fuel level of the two inside carburettors, the outside carburettor base may be used as a reference point for the measuring gauge.

2 Adjust the throttle cables by starting with the opening cable first. Loosen the locknut on the throttle opening cable, and use the adjuster to take up any slack in the cable before securing the locknut again. Loosen the locknut on the closing cable, and adjust it so there is about 2 mm ($\frac{1}{16}$ inch) of play in the throttle grip, then secure the locknut.

3 Perform the following tasks as a prelude to the actual adjustment of the carburettors at any time they are rebuilt or replaced, and especially if the engine idles roughly.

4 Remove the carburettors from the machine, if this has not already been done. Slacken and remove the three screws which retain the top cover on each carburettor, and lift the cover away. It will be noted that each of the operating rockers has an adjuster and locknut to enable the four throttle valves to be synchronised. Obtain a piece of wire approximately 0.5 - 1.0 mm (0.02 - 0.04 in) in diameter. Lift the throttle valve of one carburettor and insert the end of the wire, trapping it between the throat of the carburettor and the valve edge. Slacken the locknut of the operating rocker of the carburettor concerned, then arrange the instrument so that the mouth, and consequently the wire, points downwards.

5 Gradually screw in the adjuster until the wire **just** drops free. Note that this setting must be made with the greatest care. When set correctly, hold the adjuster in position and secure the locknut. Repeat this procedure on the remaining three instruments, after which the throttle valves of all four carburettors will be mechanically synchronised. The large knurled throttle stop screw can now be set so that the bottom edge of the throttle valves are just visible in the throat of the carburettors.

6 Before refitting the carburettors, set the pilot screws to their approximate nominal setting. Where fitted, the white plastic limiters should be pulled off. Screw in each adjuster until it seats lightly, then unscrew it by $\frac{3}{4}$ turn. Refit the limiters with the stop halfway between its normal amount of travel. Refit the carburettors, and set up the throttle cables as described in paragraph 2 of this Section.

7 Start the engine and keep it running until normal working temperature is reached, then set the throttle stop screw to achieve a tickover speed of 900 – 1200 rpm. Check that this speed remains constant after blipping the throttle once or twice. Try varying the pilot screw of each carburettor in turn. This adjustment should not exceed the range of the limiter, or $\frac{1}{4}$ turn at the most. Note the effect that this has on tickover speed, and leave it in whichever setting produces the highest speed. Reset the tickover speed to 900 – 1200 rpm, then repeat the adjustment on the remaining instruments.

7 Carburettors: synchronisation

1 In order that the carburettors be precisely synchronised it is necessary to balance them using a set of vacuum gauges connected to the inlet manifolds. To perform this operation, it will be necessary to use a set of vacuum gauges having variable restrictor, or damping, valves, such as the Kawasaki 57001-127 assembly. It will also be necessary to remove the fuel tank and to arrange a temporary remote fuel tank. For this reason it may be considered worthwhile entrusting the operation to an authorised Kawasaki Service Agent. If however, it is desired that the work be done at home, obtain the vacuum gauges, remove the petrol tank and arrange a remote supply (i.e. connect the fuel tank, placed on a nearby bench, to the carburettors, using a length of tubing).

2 Remove the small rubber caps from the inlet manifolds, and attach the vacuum gauge hoses. The vacuum gauges can now be attached, one hose to each of the four pipes, so that the vacuum on all four cylinders can all be read on the corresponding gauges. With the engine running at idle speed, close down the vacuum gauge intake valve until the gauge needle flutters less than 3 cm hg (1.2 in hg).

3 The normal manifold vacuum gauge reading is 19-24 cm Hg (7.5-9.5 cm in Hg) for each cylinder. If any gauge reads less than 2 cm Hg (0.8 in Hg) recheck the pilot air screw adjustment, also make sure that the carburettor hose clamps and sparking plugs are secure.

4 Balance the carburettors by adjusting the throttle valve adjusting screws as described in this Chapter, Section 6. All the

6.6 Splines on mixture screw head are for plastic limiter

6.7 Knurled wheel controls throttle stop on all carburettors

carburettors should be adjusted to within 2 cm Hg (0.8 in Hg) of each other.
5 Open the throttle fairly rapidly and allow it to snap shut several times, while watching to see if the vacuum gauge readings remain the same. Readjust any carburettors whose readings have changed.
6 The vacuum gauges can now be removed. Replace the protective rubber covers on the adaptors. Readjust any carburettor by the pilot screw and adjust the idle speed to about 950 – 1200 rpm.

8 Carburettors:settings

1 Some of the carburettor settings, such as the sizes of the needle jets, main jets, and needle positions are pre-determined by the manufacturer. Under normal riding conditions it is unlikely that these settings will require modification. If a change appears necessary, it is often because of an engine fault, or an alteration in the exhaust system eg; a leaky exhaust pipe connection or silencer.
2 As an approximate guide to the carburettor settings, the pilot jet controls the engine speed up to $\frac{1}{8}$ throttle. The throttle slide cut-away controls the engine speed from $\frac{1}{8}$ to $\frac{1}{4}$ throttle and the position of the needle in the slide from $\frac{1}{4}$ to $\frac{3}{4}$ throttle. The size of the main jet is responsible for engine speed at the final phase of $\frac{3}{4}$ to full throttle. These are only guide lines; there is no clearly defined demarcation line due to a certain amount of overlap that occurs.
3 Always err slightly towards a rich mixture as one that is too weak will cause the engine to overheat and burn the exhaust valves. Reference to Chapter 3 will show how the condition of the sparking plugs can be interpreted with some experience as a reliable guide to carburettor mixture strength.

9 Air cleaner:location, removal and maintenance

1 A renewable air cleaner element is mounted inside the air cleaner trunking to prevent the ingress of abrasive dust. The element may be removed for inspection and cleaning.
2 To gain access to the element, lift the dualseat to expose the top of the air cleaner chamber. Unscrew the large plastic cap and lift this away. The element can now be pulled out for examination.
3 Clean the element with petrol or a cleaning solvent and then blow it dry with compressed air from inside. Do not use any cleaner that will not completely evaporate.
4 Inspect the element and also the sponge gaskets for signs of wear or damage, and replace the element if either are damaged. The sponge gaskets can be glued back on if they are loose and in good condition. Be careful, when installing the element, not to crimp the gaskets.
5 The average useful life span of one of these elements is approximately 8000 miles or 12 months, whichever the sooner. Furthermore, if it has been cleaned three or four times due to use in very dusty conditions it should be renewed.
6 Never run the machine without the air cleaner element, otherwise the permanently weak mixture that results will cause severe engine damage.

10 Engine and gearbox lubrication

1 As previously described at the beginning of the Chapter the lubrication system is of the wet sump type, with the oil being forcibly pumped from the sump to positions at the gearbox bearings, the main engine bearings, and the cam box bearings, all oil eventually draining back to the sump. The system incorporates a gear driven oil pump, an oil filter, a safety bypass valve, and an oil pressure switch. Oil vapours created in the crankcase are vented through a breather to the air cleaner box, where they are recirculated to the crankcase, providing an oil-tight system.
2 The oil pump is of the trochoidal type, being gear driven off the crankshaft. An oil strainer is fitted to the intake side of the pump, which serves to protect the pump mechanism from impurities in the oil which might cause damage.
3 An oil filter unit is housed in the bottom of the sump, in an alloy canister containing a paper element. As the oil filter unit becomes clogged with impurities, its ability to function correctly is reduced, and if it becomes so clogged that it begins to impede the oil flow, a by-pass valve opens, and routes the oil flow around the filter. This results in unfiltered oil being circulated throughout the engine, a condition which is avoided if the filter element is changed at regular intervals.
4 The oil pressure switch which is situated inside the contact breaker housing serves to indicate when the oil pressure has dropped due to an oil pump malfunction, blockage in an oil passage, or a low oil content. The switch is not intended to be used as an indication of the correct oil level.
5 As previously mentioned an oil breather is incorporated into the system. It is mounted in the top of the crankcase, and is essential for an engine of this size with so many moving parts. It serves to minimise crankcase pressure variations due to piston and crankshaft movement, and also helps lower the oil

temperature, by venting the crankcase. The breather tube carries the crankcase vapours to the air cleaner housing where they become mixed with the air drawn into the carburettors. If the breather hose or the ports inside the breather become blocked, pressure may build up to such a level in the crankcase that oil leaks will occur. If the oil level is too high in the sump, this may result in oil misting severe enough to cause the air cleaner to become oil saturated. This will lead to poor carburation. Avoid overfilling the sump.

6 Excessive oil consumption as indicated by a blue smoke emitting from the exhaust pipes, coupled with a poor performance and fouling of the sparking plugs, is caused by either an excessive oil build up in the oil breather chamber, or by oil getting past the piston rings. First check the oil breather chamber and air cleaner for oil build up. If this is the fault, check the passageway from the air/oil separator in the oil breather chamber to the lower half of the crankcase. Blockage here will prevent oil flowing back into the crankcase, resulting in oil build-up in the breather chamber and air cleaner tube.

7 Be sure to check the oil level in the sump before starting the engine. If the oil level is not seen between the two marks adjacent to the sight 'window' at the bottom of the clutch cover, replenish with the correct amount of oil of the recommended viscosity.

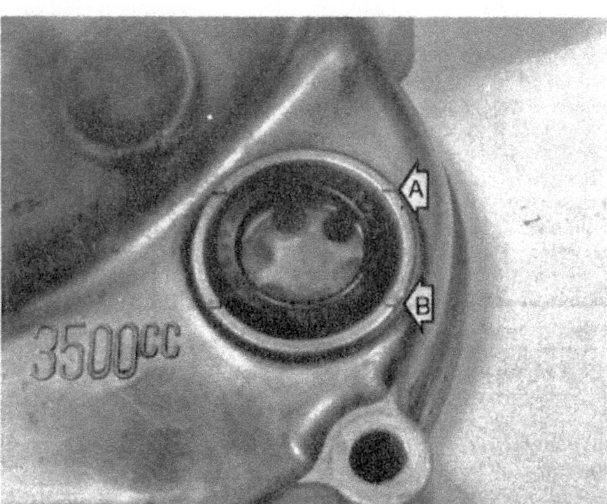

10.7 Keep oil between upper (A) and lower (B) levels

11 Lubrication system: checking the oil pressure

1 The efficiency of the lubrication system is dependent on the oil pump delivering oil at the correct pressure. This can be checked by fitting an oil pressure gauge to the right-hand oil passage plug, which is located immediately below the contact breaker housing. Note that the correct threaded adaptor must be obtained or fabricated for this purpose. The best course of action is to obtain the correct Kawasaki pressure gauge and adaptor, part numbers 57001-164 and 57001-403 respectively.

2 Remove the end plug and fit the adaptor and gauge into position. The correct pressure at 4000 rpm, 90°C (194°F) should be 2.0 - 2.5 kg/cm^2 (28-36 psi). If the oil pressure is significantly below this figure, and no obvious oil leakage is apparent, the oil pump should be removed for examination. On no account should the machine be used with low oil pressure, as plain bearing engines in particular rely on oil pressure as much as volume for effective lubrication.

3 It is likely that the normal oil pressure will be slightly above the specified pressure, but if it proves to be abnormally high, it is likely to be due to the oil pressure relief valve being jammed or damaged. The latter component is fitted to the inside of the sump. Refer to Section 13 of this Chapter.

9.3 Element may be washed in suitable solvent (see text)

12 Oil pump: removal, examination and renovation

1 It is possible to remove the oil pump for examination after the clutch cover, clutch assembly, and sump have been removed, having, of course, drained the sump beforehand. Refer to the relevant Sections of Chapter 1 for details. Examination of the pressure relief valve and renewal of the oil filter element should be undertaken before reassembling these components. The oil pump can be removed from the underside of the crankcase after releasing the mounting bolts.

2 The oil pump end cover is retained by three screws. These should be removed. and the circlip on the end of the oil pump spindle detached, to permit the cover to be lifted away. The inner and outer rotors can be shaken out of the pump body, the driving pin displaced, and the pump spindle withdrawn.

3 Wash all the pump components with petrol and allow them to dry before carrying out an examination. Before partially reassembling the pump for various measurements to be carried out, check the casting for breakage or fracture, or scoring on the inside perimeter.

4 Reassemble the pump rotors and measure the clearance between the outer rotor and the pump body, using a feeler gauge. If the measurement exceeds the service limit of 0.30 mm (0.012 in) the rotor or the body must be renewed. whichever is worn. Measure the clearance between the outer rotor and the inner rotor, using a feeler gauge. If the clearance exceeds 0.30 mm (0.012 in) the rotors must be renewed as a set. It should be noted that one face of both the inner and the outer rotor is punch marked. The punch marks should face away from the main pump casing during measurements and on reassembly. With the pump rotors installed in the pump body lay a straight edge across the mating surface of the pump body. Again with a feeler gauge measure the clearance between the rotor faces and the straight edge. If the clearance exceeds 0.12mm (0.005 in) the rotors should be replaced as a set.

5 Examine the rotors and the pump body for signs of scoring, chipping or other surface damage which will occur if metallic particles find their way into the oil pump assembly. Renewal of the affected parts is the only remedy under these circumstances, bearing in mind that the rotors must always be replaced as a matched pair.

6 Reassemble the pump components by reversing the dismantling procedure. The component parts must be ABSOLUTELY clean or damage to the pump will result. Replace the rotors and lubricate them thoroughly before refitting the cover.

7 Check that the pump turns smoothly, then refit it to the casing. Before refitting the sump, remove and examine the pressure relief valve as described in Section 13. Do not omit the large 'O' ring which must be fitted to the oil pump before the sump is refitted.

13 Oil pressure relief valve: dismantling, examination and renovation

1 If problems with the lubrication system have been experienced, it is advisable to check the operation of the pressure relief valve, whilst the sump is removed. The valve can be unscrewed from the inside of the sump, using a spanner on the hexagonal body.

2 Remove the small internal circlip from the top of the valve body to release the internal parts. On some early models, a spring-loaded ball-type valve was used, whereas on later models, a small plunger replaces the ball. The operation and method of checking, however, remains the same for both types.

3 Examine the plunger (or ball) for signs of wear or scoring, renewing it, if necessary. The bore of the valve body should also be free from scoring. If the valve appears badly worn, the complete assembly should be renewed. Renew the spring if it has a free length of less than 19.1 mm (0.75 in).

12.1 Oil pump can be removed with engine in position

12.2a Remove circlip and screws to release pump end plate

12.2b Shake out the outer and inner rotors for examination

12.2c Measure the clearance between rotors and pump body

12.2d Displace the driving pin to allow ...

12.2e ... the driving gear and shaft to be withdrawn

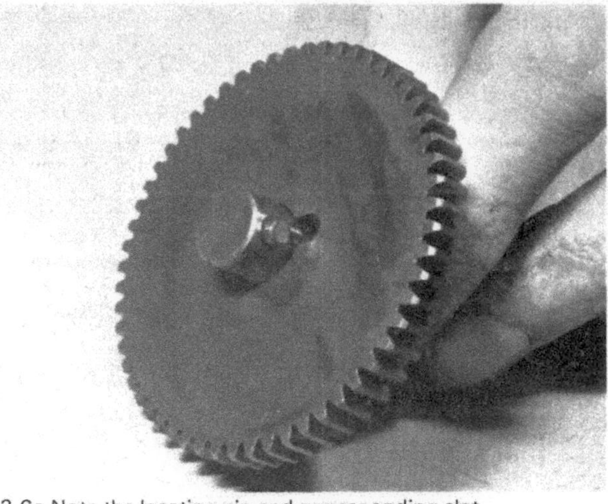
12.6a Note the locating pin and corresponding slot

12.6b Use new gasket, note locating dowel

12.6c Filter screen can be removed to permit cleaning

13.1 Pressure relief valve is screwed into sump

14 Oil filter: renewing the element

1 The oil filter is contained within a semi-isolated chamber within the crankcase. Access to the element is made by unscrewing the filter cover centre bolt which will bring with it the cover and also the element. Before removing the cover place a receptacle beneath the engine to catch the engine oil contained in the filter chamber.

2 When renewing the filter element it is wise to renew the filter cover 'O' ring at the same time. This will obviate the possibility of any oil leaks. Do not overtighten the centre bolt on replacement.

3 The filter by-pass valve, comprising a plunger and spring, is situated in the bore of the filter cover centre bolt. It is recommended that the by-pass valve be checked for free movement during every filter change. The spring and plunger are retained by a pin across the centre bolt. Knocking the pin out will allow the spring and plunger to be removed for cleaning.

4 Never run the engine without the filter element or increase the period between the recommended oil changes or oil filter changes.

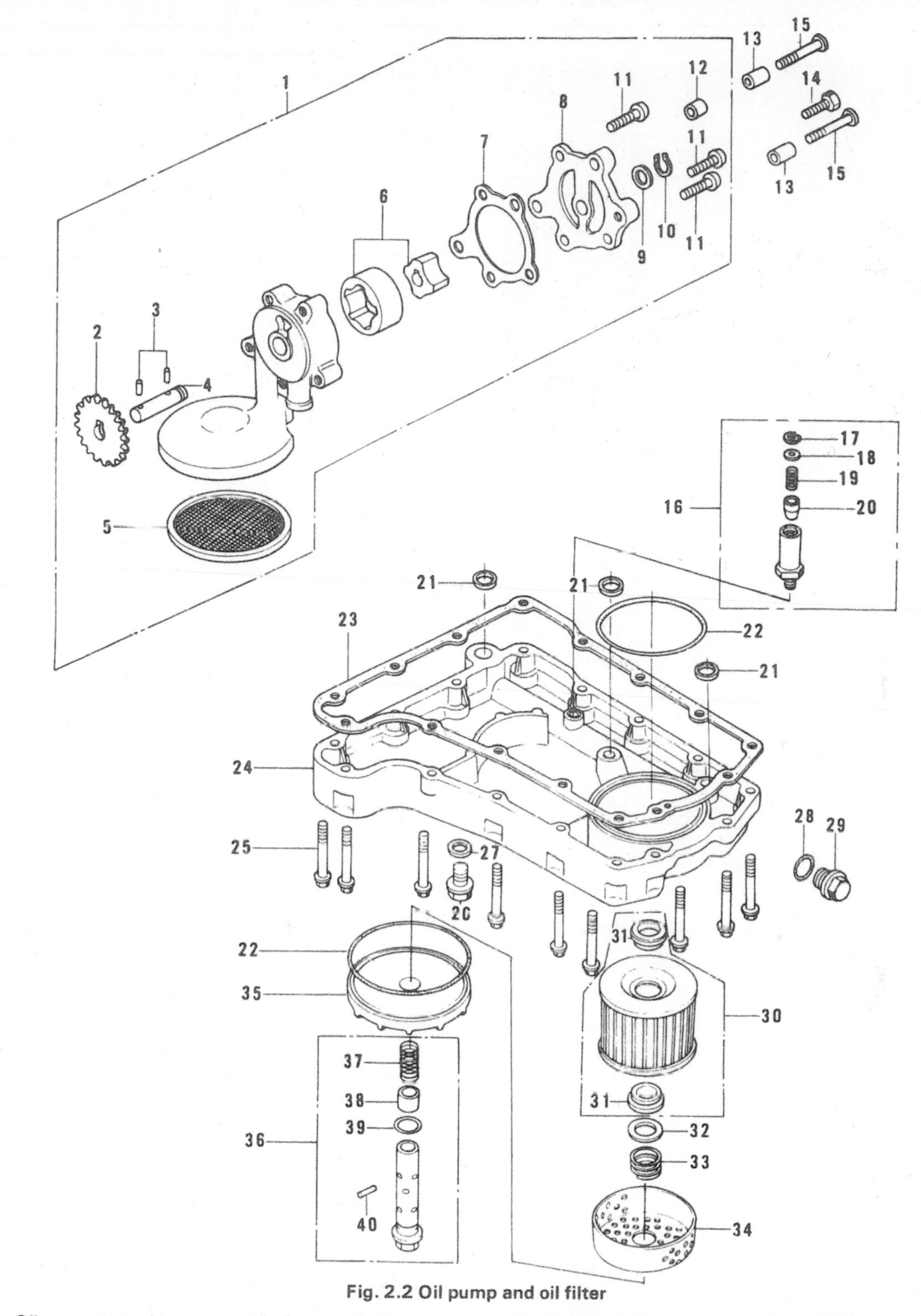

Fig. 2.2 Oil pump and oil filter

1 *Oil pump assembly*
2 *Oil pump gear*
3 *Dowel pin – 2 off*
4 *Oil pump shaft*
5 *Oil pump gauze strainer*
6 *Oil pump rotor assembly*
7 *Oil pump gasket*
8 *Oil pump cover*
9 *Washer*
10 *Circlip*
11 *Screw – 3 off*
12 *Dowel pin*
13 *Dowel pin – 2 off*
14 *Bolt*
15 *Screw – 2 off*
16 *Relief valve assembly*
17 *Circlip*
18 *Spacer*
19 *Spring*
20 *Piston*
21 *O ring – 3 off*
22 *O ring – 2 off*
23 *Oil sump gasket*
24 *Oil sump cover*
25 *Bolt – 15 off*
26 *Oil drain plug*
27 *Washer*
28 *O ring*
29 *Plug*
30 *Oil filter*
31 *Oil filter grommet*
32 *Washer*
33 *Oil filter spring*
34 *Oil filter plate*
35 *Oil filter cover*
36 *Oil filter bolt assembly*
37 *Valve spring*
38 *Bypass valve*
39 *O ring*
40 *Spring pin*

14.1 Oil filter is retained by central bolt

15 Oil pressure warning switch

1 An oil pressure warning switch is incorporated in the lubrication system to give warning of impending disaster in the event of oil pressure failure. The switch is located inside the contact breaker housing and normally gives very little trouble. In the event that the oil warning light does not come on when the ignition is just switched on, it is imperative that the fault is isolated and rectified before the machine is ridden.

2 Remove the contact breaker cover, and disconnect the pressure switch lead. With the ignition switched on, earth the lead from the warning lamp against the crankcase. If the warning lamp comes on, the switch should be renewed. If, however, the warning lamp still does not work, attention should be turned to the bulb and wiring.

3 The switch may be unscrewed from the casing after the contact breaker base plate has been removed, noting that it will be necessary to re-time the ignition after reassembly. Note that the terminal which screws onto the top of the switch should be positioned away from the contact breaker to avoid any possibility of it earthing across. This problem actually occurred, when the machine used for the photographs in this manual was being road tested.

If the light comes on suddenly whilst riding, stop the machine immediately, and investigate the cause, noting the above comments. On no account ride the machine with the warning lamp on.

15.1 Oil pressure switch is mounted inside contact breaker housing

16 Fault diagnosis: fuel system and lubrication

Symptom	Cause	Remedy
Engine gradually fades and stops	Fuel starvation	Check vent hole in filler cap. Sediment in filter bowl or float chamber. Dismantle and clean.
Engine runs badly. Black smoke from exhausts.	Carburettor flooding	Dismantle and check carburettor. Check for punctured float or sticking float needle.
Engine lacks response and overheats	Weak mixture	Check for partial block in carburettors
	Air cleaner disconnected or hose split	Reconnect or renew hose.
	Modified silencer has upset carburation	Replace with original design
Oil pressure warning light comes on	Lubrication system failure	Stop engine immediately. Trace and rectify fault before re-starting.
	Short circuit in warning lamp system	Check and rectify source of short (see text).
Engine suddenly gets noisy	Failure to change engine oil when recommended.	Drain off old oil and refill with new oil of correct grade. Renew oil filter element.

Chapter 3 Ignition system

Contents

Specifications

Sparking plugs:

Size	14 mm
Reach	19 mm ($\frac{3}{4}$ in)
*Type (Normal use)	NGK B8ES or ND W24ES Champion N4
(Low speed use and running in)	NGK B7ES
(High speed use and racing)	NGK B9ES
Alternative:	Motorcraft AG1 or AG901 (high speed)
Electrode gap:	0.7 - 0.8 mm (0.028 - 0.031 in)

*Recommended by manufacturer

Condenser

Capacity:	0.24 ± 0.02 microfarad

Ignition coils:

Number:	2
Type:	ZC002-14, ZC002-23
Make:	Toyo Denso
Primary winding resistance:	4.0 ohms approx.
Secondary winding resistance:	23.0 k ohms approx.

Contact breakers

Gap:	0.3 - 0.4 mm
Dwell angle (degrees)	185° - 195°
Dwell angle (percentage)	51 - 54%
Ignition timing	
Range:	10° BTDC @ 1600 rpm to 35° BTDC @ 3200 rpm
Advance method:	Mechanical, by Automatic Timing Unit (ATU)

1 General description

The spark necessary to ignite the petrol vapour in the combustion chambers is supplied by a battery and two ignition coils (one coil to two cylinders).

There are two sets of contact breaker points, two condensers, four sparking plugs and an automatic ignition advance mechanism. The breaker cam, which is incorporated in the advance mechanism, opens each set of points once in 180° of crankshaft rotation, causing a spark to occur in two of the cylinders. The other set of points fires 180° later, so that in every 360° of crankshaft rotation each plug is fired once. One extra spark occurs during the time when there is no combustible material in the chamber.

Each set of points has one fixed and one movable contact, the latter of which pivots as the lobe of the cam separates them. The two condensers are wired in parallel, one with each set of contact points, and these function as electrical storage reservoirs, whilst also preventing arcing across the points. The condensers serve to absorb surplus current that tries to run back through the system where there is an overload situation, and feeds the current back to the ignition coils. They also help intensify the spark. When the points are closed, the current flows straight through them to earth. When they open, there is now an open circuit. If not for the condensers, the current may arc across the points causing them to burn and pit. When the condensers reach their capacity, they discharge the current back through the primary windings and eventually to the sparking plug. Any time the points get badly burnt, it is advisable to renew them, and the condensers also.

Each of the two coils has two high voltage sparking plug leads, and as in the case of points, one coil serves cylinders 1 and 4, and the other, cylinders 2 and 3.

The coils convert the low tension voltage into a high tension voltage sufficient to provide a spark strong enough to jump the sparking plug air gap., If at any time a very weak or erratic spark occurs at the plug, and the rest of the ignition system is known to be in good condition, it is time to renew an ignition coil. Although coils normally have a long life they can sometimes be faulty, especially if the outer case has been damaged.

The automatic advance mechanism serves to advance the ignition timing as the engine rpm rises. The mechanism is made up of two spring loaded weights which, under the action of centrifugal force created by the rotation of the crankshaft, fly apart and cause the contact points to open earlier. If the mechanism does not operate smoothly, the timing will not advance smoothly, or it may stick in one position. This will result in poor running in any but that one position. Sometimes the springs are prone to stretching, which can cause the timing to advance too soon. It is best to check the automatic advance mechanism, by carrying out a static timing test on the ignition followed by a strobe test. It is always best to check the motion of the weights by hand every 2000 miles and to clean and lubricate the unit at the same time.

The ignition system is operated by a key switch, mounted on a dash panel between the speedometer and the tachometer. There are three positions on the switch, OFF, ON, and PARK. In the OFF position all the circuits are turned off and the key can be removed from the switch. In the ON position the motorcycle can be started and all the lights and accessories can be used. The key cannot be removed from the switch when it is in this position.

In the PARK position, only the tail light stays on, and the key can be removed from the switch. The charging of the battery that operates the ignition system is taken care of by an ac alternator that is mounted on the left-hand side of the crankshaft. This supplies current which is rectified by a rectifier, mounted on a panel alongside the voltage regulator, on the right-hand side of the machine, below the dualseat.

2 Ignition coils: checking

1 The ignition coils are a sealed unit designed to give long life, and are mounted on the frame tubes in the upper cradle behind the steering head. The most accurate test of an ignition coil is with a three point coil and condenser tester (electrotester).

2 Connect the coil to the tester when the unit is switched on, and open out the adjusting screw on the tester to 6 mm (0.24 inch). The spark at this point should bridge the gap continuously. If the spark starts to break down or is intermittent, the coil is faulty and should be renewed.

3 In the absence of a coil tester, the winding may be checked for broken or shorted windings using a multimeter, noting that test will not reveal insulation breakdown which may only be evident under high voltage.

4 The primary winding resistance can be measured by connecting one of the meter probes to the red/yellow lead, and the other to the green or black lead. Resistance should be in the region of 4 ohms. Secondary resistance involves the connection of one probe to each of the high tension leads. In this case the resistance should be approximately 23 k ohms. Finally check for any conductivity between the coil body and the red/yellow lead. If the insulation here is not perfect, or if the primary and secondary resistance readings are noticeably different from those specified, then the coil should be renewed.

3 Contact breaker adjustments

1 To gain access to the contact breaker it is necessary to remove the cover plate screws and the cover on the right-hand front of the crankcase.

2 Rotate the engine by slowly turning it over with the kickstarter until one set of points is fully open. Examine the faces of the contacts for pitting and burning. If badly pitted or burnt they should be renewed as described in Section 4 of this Chapter.

3 Adjustment is carried out by slackening the screws on the base of the fixed contact, and adjusting the gap within the range 0.3 to 0.4 mm (0.012 - 0.016 inch) when the points are fully open by moving the base contact with a screwdriver at the slotted point. Retighten the two screws after adjustment with the feeler gauge and re-check the gap, then repeat the same operation for the other set of points. Do not forget to double check after you have tightened the setting screws, in case the setting has altered.

4 Before replacing the cover and gasket, place a slight smear of grease on the cam and a few drops of oil on the felt pad. Do not over lubricate for fear of oil getting on the points, and causing poor electrical contact.

2.1 Ignition coils are mounted on lugs beneath petrol tank

3.2 Use feeler gauge to check contact breaker gap

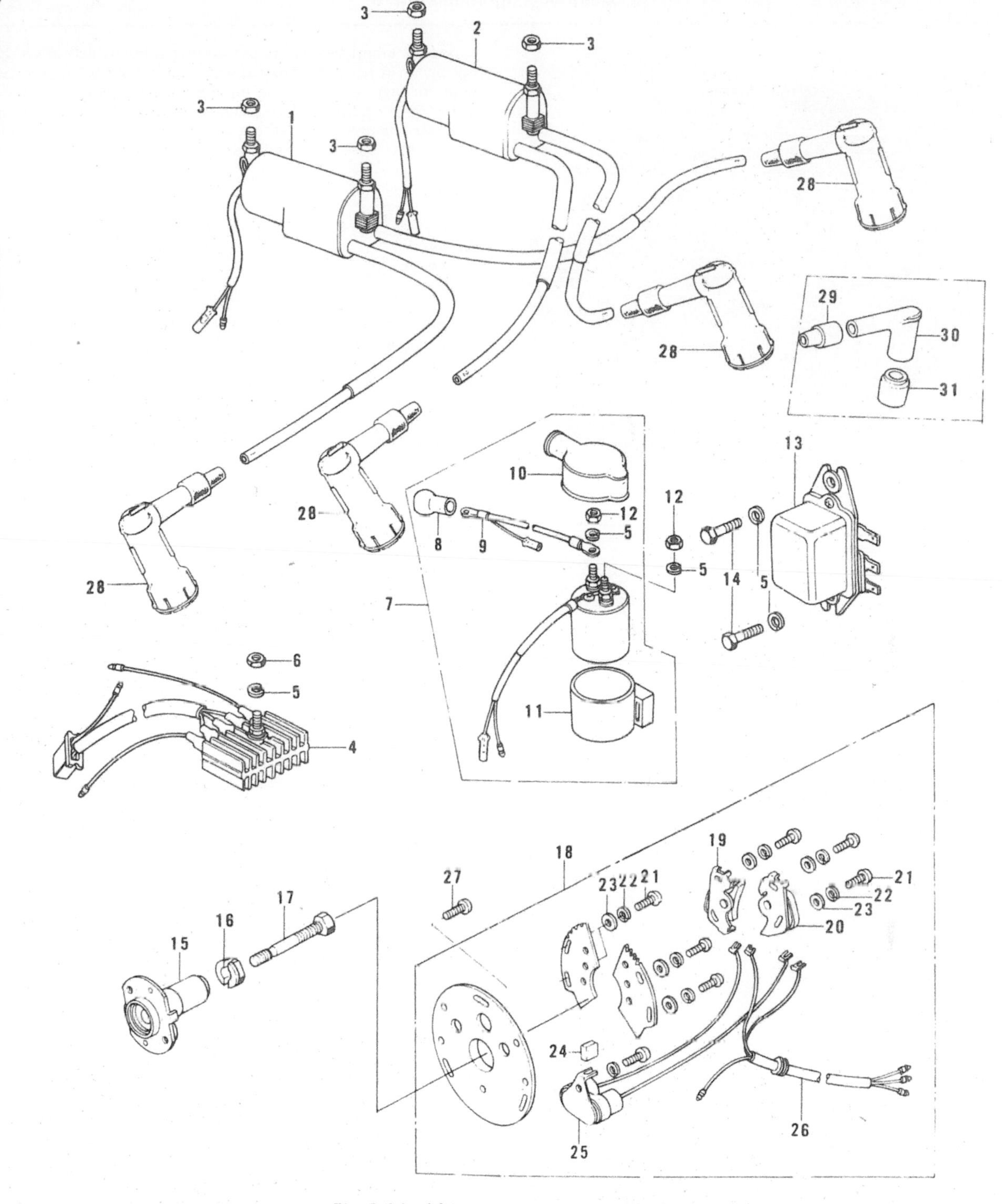

Fig. 3.1 Ignition system – component parts

1 *Ignition coil, outer cylinders*
2 *Ignition coil, inner cylinders*
3 *Nut – 4 off*
4 *Rectifier*
5 *Spring washer – 5 off*
6 *Nut*
7 *Starter solenoid*
8 *Dust cap*
9 *Solenoid lead*
10 *Rubber shroud*
11 *Rubber mounting block*
12 *Nut – 2 off*
13 *Regulator*
14 *Bolt – 2 off*
15 *Automatic timing unit (ATU)*
16 *Hexagon*
17 *Special bolt*
18 *Contact breaker assembly*
19 *Contact breaker set, inner cylinders*
20 *Contact breaker set, outer cylinders*
21 *Screw – 9 off*
22 *Spring washer – 9 off*
23 *Plain washer – 9 off*
24 *Lubricating wick*
25 *Condenser assembly*
26 *Contact breaker wiring*
27 *Screw – 3 off*
28 *Suppressor cap – 4 off*
29 *Dust seal – 4 off (alternative part)*
30 *Suppressor cap – 4 off (alternative part)*
31 *Dust seal – 4 off (alternative part)*

3.4 Lubricate felt wick (arrowed) with one or two drops of oil

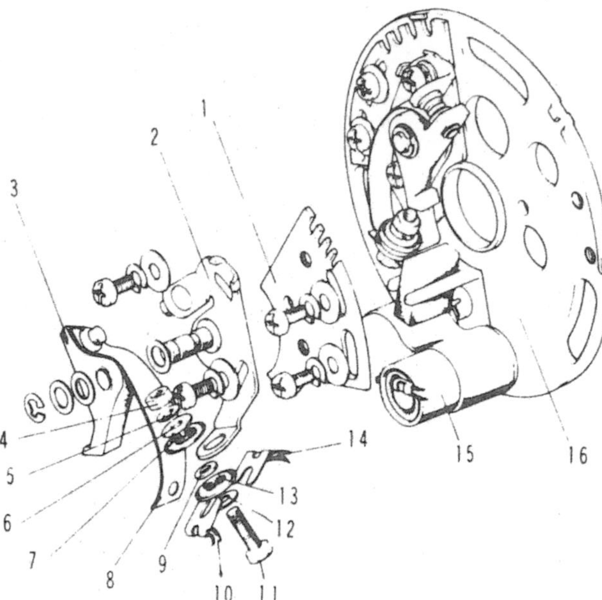

Fig. 3.2 Contact breaker assembly

1 Contact breaker base plate segment – 2 off
2 Fixed contact – 2 off
3 Moving contact – 2 off
4 Nut – 2 off
5 Spring washer – 2 off
6 Plain washer – 2 off
7 Insulating washer – 2 off
8 Leaf spring – 2 off
9 Insulating washer – 2 off
10 Contact breaker terminal – 2 off
11 Terminal bolt – 2 off
12 Plain washer – 2 off
13 Insulating washer – 2 off
14 Condenser terminal – 2 off
15 Condenser – 2 off
16 Contact breaker base plate

4 Contact breaker adjustment – using dwell meter

1 The contact breaker gap may be set by using a dwell meter. These are obtainable, usually in the form of a combined test meter or ignition analyser, from most Auto Accessory Dealers. Setting the contact breakers by this method ensures that the contact breakers produce the performance which is normally expected when the gap is set correctly, but which often does not occur in practice. Most dwell meters have a range of settings for engines of varying numbers of cylinders, and may be calibrated in degrees of crankshaft rotation or in percentage. The table below shows the various cylinder number settings, and the corresponding readings:

Setting	*Reading*
1 cylinder	*185.0 - 195.0° (51.0 - 54.0%)*
2 cylinder	*92.5 - 97.5° (22.5 - 27.0%)*
3 cylinder	*62.5 - 65.0° (17.0 - 18.0%)*
4 cylinder	*46.5 - 49.0° (13.0 - 13.5%)*

2 Set the dwell tester in the appropriate positions, then connect the positive (+) probe to the contact breaker terminal, and the negative (-) probe to the crankcase. With the engine running at less than 1050 rpm, check that the reading is within the limits given. If this is not the case, slacken the fixed contact securing screw just enough to permit movement, then adjust the gap until the reading is within the specified tolerances. Tighten the securing screw, then check that the setting has not altered. The test sequence should be repeated on the remaining set of contact breaker points.

5 Contact breaker points: removal and replacement

1 If the contact points are badly burnt or worn, they should be renewed. Undo the two screws that hold the base of the fixed contact of each set of points, and remove the wire leading to the condenser, which will allow the points to be lifted off. Removal of the circlip on the end of the pivot pin will permit the moving contact point to be detached. Note the arrangement of the insulating washers.
2 The points should be dressed with an oil stone or fine emery cloth to remove deposits due to arcing. Keep them absolutely square throughout the dressing operation, otherwise they will make angular contact on reassembly, and rapidly burn away. If emery cloth is used, it should be backed by a flat strip of steel. If it is necessary to remove a substantial amount of material before the faces can be restored, the points should be renewed.
3 Replace the contacts by reversing the dismantling procedure, making quite certain that the insulating washers are fitted in the correct way. In order for the ignition system to function at all, the moving contact and the low-tension lead must be perfectly insulated from the base plate and fixed contact. It is advantageous to apply a very light smear of grease to the pivot pin, prior to replacement of the moving contact.
4 Check, and if necessary, re-adjust the contact breaker points when they are in the fully open position.

6 Condensers: removal and replacement

1 There are two condensers contained in the ignition system, each one wired in parallel with a set of points. If a fault develops in a condenser, ignition failure is likely to occur.
2 If the engine proves difficult to start, or misfiring occurs, it is possible that a condenser is at fault. To check, separate the contact points by hand when the ignition is switched on. If a spark occurs across the points as they are separated by hand and they have a burnt or blackened appearance, the condenser connected to that set of points can be regarded as unserviceable.
3 Test the condenser on a coil and condenser tester unit or alternatively fit a new replacement. In view of the small cost involved it is preferable to fit a new condenser, and observe the effect on engine performance as a result of the substitution.
4 Check that the screws that hold the condensers to the contact breaker plate are tight, and also form a good earth connection.

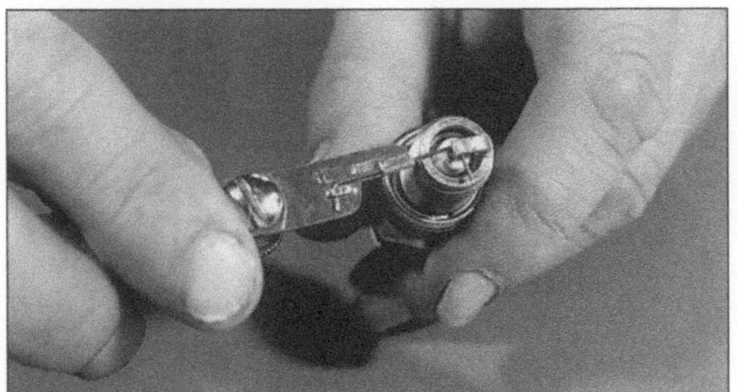
Electrode gap check – use a wire type gauge for best results.

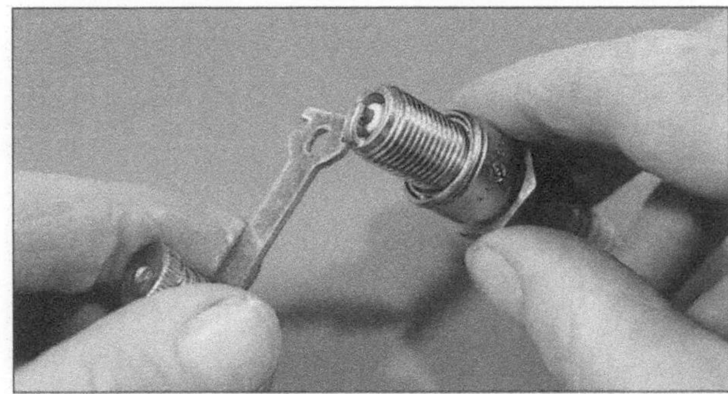
Electrode gap adjustment – bend the side electrode using the correct tool.

Normal condition – A brown, tan or grey firing end indicates that the engine is in good condition and that the plug type is correct.

Ash deposits – Light brown deposits encrusted on the electrodes and insulator, leading to misfire and hesitation. Caused by excessive amounts of oil in the combustion chamber or poor quality fuel/oil.

Carbon fouling – Dry, black sooty deposits leading to misfire and weak spark. Caused by an over-rich fuel/air mixture, faulty choke operation or blocked air filter.

Oil fouling – Wet oily deposits leading to misfire and weak spark. Caused by oil leakage past piston rings or valve guides (4-stroke engine), or excess lubricant (2-stroke engine).

Overheating – A blistered white insulator and glazed electrodes. Caused by ignition system fault, incorrect fuel, or cooling system fault.

Worn plug – Worn electrodes will cause poor starting in damp or cold conditions and will also waste fuel.

6.1 Condensers are mounted on contact breaker plate

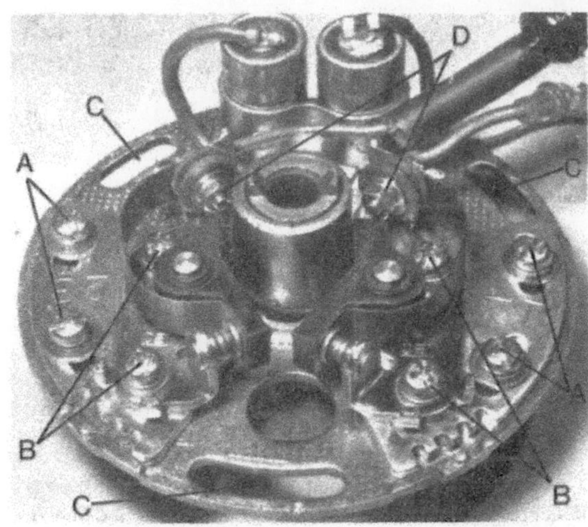

7.4 A: Contact breaker segment screws. B: Fixed contact screws C: Base plate screw slots. D: Contact breaker terminals

7 Ignition timing: checking and resetting

1 In order to check the ignition timing statically it is necessary to remove the contact breaker cover, and check the point gaps first. Using a feeler gauge, set the gaps within the range of 0.3 - 0.4 mm (0.012 - 0.016 in) when each set of points is fully open.
2 Rotate the engine with a spanner on the timing shaft bolt, until the 'F' mark on the timing advancer for the set of points to be adjusted is slightly to the left of the timing mark, located just above the timing advancer.
3 Make up a timing light (a tail light bulb and two pieces of wire; one wire to the base of the bulb and one wire to the side of the bulb). Fit crocodile clips to the other ends of the wire, and connect one wire to the rocker arm point spring, and the other to earth (an engine cooling fin or the frame etc) by means of the crocodile clips. Turn on the ignition switch, rotate the engine backwards a few degrees past the point of alignment, and then rotate forwards again until the 'F' mark is aligned with the timing mark. The light should flicker as the two come together. If it does, the timing is correct. Check in a similar fashion the other set of points.
4 If the mounting plate for the points has to be moved, loosen the two screws on the point adjusting plate and use a screwdriver to lever the plate in the position desired, so that the points just begin to open as the timing marks come into line. If the points adjusting plate will not travel far enough for any set of points, loosen the contact breaker base plate by unscrewing the three crosshead screws enough to rotate the baseplate to the required position. Secure the screws and recheck the timing before going on to the other set of points. **Note:** Do not leave the ignition switched on for long, just enough to set the timing, otherwise the points will burn.
5 If the light will not come on, there could be a bad connection, the points gap may need attention again, or the points may need cleaning thoroughly.
6 It cannot be overstressed that optimum performance depends on the accuracy with which the ignition timing is set. Even a small error can cause a marked reduction in performance and in an extreme case, engine damage due to overheating. The contact breaker gap must be checked and if necessary re-adjusted **BEFORE** carrying out ignition timing checking or setting. When ignition timing evaluation is made it is essential that the automatic timing remains in the retarded (closed) position.

8 Automatic timing unit:examination

1 The automatic timing mechanism rarely requires attention, although it is advisable to examine it periodically, when the contact breaker is receiving attention. It is retained by a small bolt and washer through the centre of the integral contact breaker cam and can be pulled off the end of the camshaft when the contact breaker plate is removed.
2 The unit comprises a spring loaded balance weights, which move outward against the spring tension as centrifugal force increases. The balance weights must move freely on their pivots and be rust-free. The tension springs must also be in good condition. Keep the pivots lubricated and make sure the balance weights move easily, without binding. Most problems arise as a result of condensation, within the engine, which causes the unit to rust and balance weight movement to be restricted.
3 The automatic timing unit mechanism is fixed in relation to the crankshaft by means of a dowel. In consequence the mechanism cannot be replaced in anything other than the correct position. This ensures accuracy of ignition timing to within close limits, although a check should always be made when reassembly of the contact breakers is complete.
4 The correct functioning of the auto-advance unit can be checked when the engine is running by the use of a stroboscopic light. If a strobe light is available, connect it to the ignition circuit as directed by the manufacturer of the light. With the engine running, direct the beam of light at the fixed timing mark on the crankcase, through the aperture in the base plate. At tickover (1100 - 1300 rpm) the timing mark and the 'F' mark on the auto-advance unit should be precisely aligned. When the engine is running at 3000 rpm or above, the timing mark should align with two parallel lines which are marked on the automatic timing unit slightly in advance of the 'F' mark. The above test relies, of course, on the static ignition timing being correct.

9 Sparking plugs: checking and resetting the gaps

1 Four NGK B8ES, ND W24ES or Champion N4 sparking plugs are fitted as standard to the Kawasaki Z650 series. Under certain operating conditions, a change from these plug grades may be required, but generally the type recommended above will be found to give the best results. If the plugs persistently

become fouled, it may be necessary to substitute a grade of sparking plug which operates at a higher temperature, such as NGK B7ES. Conversely, prolonged high speed use may result in the standard sparking plugs becoming burnt, in which case NGK B9ES or equivalent may be used.

2 The sparking plug electrode gaps should be checked in accordance with the intervals specified in the Routine Maintenance Section, or in the event of ignition problems. Note that if the sparking plugs in 1 and 4 or 2 and 3 cylinders appear to malfunction simultaneously, the relevant coil should be checked (See Section 2 of this Chapter).

To reset the gap, bend the outer electrode to bring it closer to, or further away from the central electrode until a 0.7 mm (0.028 in) feeler gauge can be inserted. Never bend the central electrode or the insulator will crack, causing engine damage if the particles fall into the cylinder whilst the engine is running.

3 With some experience, the condition of the sparking plug electrodes and insulator can be used as a reliable guide to engine operating conditions. See the accompanying diagram.

4 Always carry a spare pair of sparking plugs of the recommended grade. In the rare event of plug failure, they will enable the engine to be restarted.

5 Beware of over-tightening the sparking plugs, otherwise there is risk of stripping the threads from the aluminium alloy cylinder heads. The plugs should be sufficiently tight to seat firmly on their copper sealing washers, and no more. Use a spanner which is a good fit to prevent the spanner from slipping and breaking the insulator.

6 If the threads in the cylinder head strip as a result of overtightening the sparking plugs, it is possible to reclaim the head by the use of a Helicoil thread insert. This is a cheap and convenient method of replacing the threads; most motorcycle dealers operate a service of this nature at an economic price.

7 Make sure the plug insulating caps are a good fit and have their rubber seals. They should also be kept clean to prevent tracking. These caps contain the suppressors that eliminate both radio and TV interference.

10 Ignition switch – maintenance

1 In the event that the ignition switch malfunctions, it may be checked by using a multimeter set on resistance, or alternatively a simple battery/bulb test arrangement, such as that described in Section 7, may be used.

2 Remove the headlamp unit, trace and separate the connector block from the ignition switch, and test the continuity between the various leads, using the table below as a guide:

Ignition switch 'ON'		*Result*
Check :	*White to Brown*	*Continuity*
	Blue to Red	*Continuity*
	White to Red	*Isolation*
	Brown to Blue	
	White to Blue	
	Brown to Red	

Ignition switch 'PARK'		
Check :	*White to Red*	*Continuity*
	White to Brown	*Isolation*
	White to Blue	
	Brown to Blue	
	Brown to Red	

3 If any of the results are not as shown above, the switch must be regarded as defective. It may be possible to restore the contacts by using a proprietary switch cleaning fluid, but if this fails, a new switch will be required.

8.1 Automatic timing unit is retained by centre bolt

8.2 Check the action of springs, weights and cam

11 Fault diagnosis: ignition system

Symptom	Cause	Remedy
Engine will not start	Faulty ignition switch	Operate switch several times in case contacts are dirty. If lights etc, function, switch may need renewal.
	Starter motor not working	Discharged battery. Use kickstart until battery is recharged.
	Short circuit in wiring	Check whether fuse is intact. Eliminate fault before switching on again.
	Completely discharged battery	If lights do not work, remove battery and recharge.
Engine misfires	Faulty condenser in ignition circuit	Renew condenser and re-test.
	Fouled sparking plug	Renew plug and have original cleaned.
	Poor spark due to generator failure and discharged battery	Check output from generator. Remove and recharge battery
Engine lacks power and overheats	Retarded ignition timing	Check timing and also contact breaker gap. Check whether auto-timing unit has jammed.
Engine 'fades' when under load, rattle from engine	Pre-ignition	Check grade of plugs fitted; use recommended grades only.

Chapter 4 Frame and forks

Contents

Specifications

Frame

Type	Tubular, double cradle

Front forks

Type	Hydraulically damped telescopic
Damping oil capacity	183 – 191 cc per leg, dry
Damping oil grade	SAE 15W
Damping oil level	385 mm (15·16 in) from top of stanchion
Fork spring free length	444·4 mm nominal (17·5 in)
	456·7 mm wear limit (17·9 in)

Rear suspension units

Type	Coil spring, hydraulically damped

Swinging arm

Type	Welded tubular steel
Pivot sleeve diameter	21·979 – 22·000 mm (0·865 – 0·866 in)
Wear limit	21·950 mm (0·864 in)
Bush internal diameter	22·055 – 22·088 mm (0·868 – 0·870 in)
Wear limit	22·29 mm (0·878 in)
Pivot shaft runout	less than 0·1 mm (0·004 mm)
Wear limit	less than 0·2 mm (0·008 in)

1 General description

1 The frame of the Kawasaki Z650 is of the full cradle type, in which the engine is supported by duplex tubes at the base of the crankcase.

2 A top tube runs from the steering head to a position at the rear of the petrol tank; the frame is extended to the rear mudguard with provision for fitting the mudguard. Lugs for the attachment of the dualseat, the pillion footrests, rear brake pedal, centre stand and prop stand are fitted to the frame.

3 The front forks are hydraulically damped, consisting of two telescopic shock absorber assemblies, each of which comprises an inner tube, an outer tube, a spring and a cylinder, piston and valve. The whole fork assembly is attached to the frame by the steering head stem and is mounted on two bearing assemblies contained in the steering head housing.

4 The damping action of the fork is accomplished by the flow resistance of the fork oil flowing between the inner and outer tubes. The method of removal, dismantling, and reassembly of the complete fork assembly is described in the following text.

2 Front fork legs: removal and replacement

1 The front fork legs may be removed from the machine without disturbing the fork yokes, avoiding the complication of removing the handlebar components and headlamp assembly. Start by supporting the machine securely on its centre stand. Place a stout wooden box or similar, underneath the crankcase to raise the front wheel clear of the ground.
2 Disconnect the speedometer cable at the wheel by releasing the gland nut which retains it to the speedometer drive unit. Slacken the two nuts which retain each wheel spindle clamp, then slacken the wheel spindle sleeve nuts. The clamps may now be removed completely, and the wheel lifted clear of the forks. Note that there is a risk of the brake pads being displaced if the lever should be accidentally squeezed while the wheel is removed. To prevent this occurring, a small chip of plywood or similar should be inserted between the brake pads.
3 If the fork leg which carries the brake caliper is to be removed, the caliper must be detached. This latter unit is retained by two mounting bolts which pass through lugs on the lower leg. Remove the caliper and release the hydraulic hose guides, then tie the caliper away from the fork legs. Do not allow it to hang from its hydraulic hose. Remove the six front mudguard mounting bolts, and lift the mudguard away. Place this in a safe place where the finish will not get scratched or chipped.
4 The fork legs are each retained by a clamp bolt in the lower fork yoke, and a similar clamp arrangement on the upper yoke. Release the relevant clamp bolts. The complete fork leg(s) can now be pulled downwards and disengaged from the yokes, leaving the yokes and headlamp assembly in position.
5 The fork legs can be refitted in a similar manner, twisting the stanchion to and fro as it is slid into position. If it proves difficult to fit the stanchions through the yokes, the clamp bolts may be removed and a screwdriver blade used to open the holes in the yokes slightly.
6 The top nuts should be just flush with the top of the upper yoke before retightening the clamp bolts. Remember to refill the fork legs if they have been drained. An empty gear oil squeeze pack can be used to top up the legs with SAE 10W/20 engine oil. The correct level is 385 mm (15-16 in) from the top of the stanchion.

3 Fork yokes and steering head bearings: removal and replacement

1 The fork yokes and steering head assembly can be removed with or without the fork legs in position. Start by following paragraphs 1 to 3 inclusive in Section 2 of this Chapter. Unlatch the seat and disconnect the petrol feed pipe, then disengage and remove the petrol tank, placing it somewhere where it will not get damaged. Remove the headlamp unit, and disconnect the various leads and connectors, pulling the cables clear of the headlamp shell. Remove the headlamp shell and indicator lamps, then disconnect the speedometer and tachometer drive cables, pulling these clear of the fork assembly.
2 Release the instrument panel mounting nuts, taking care not to lose the mounting rubbers. The complete assembly can now be lifted away from the top fork yoke and placed to one side. Disconnect the front brake switch leads, thus releasing the three-way hydraulic union from the lower fork yoke.
3 Slacken the four handlebar clamp bolts and remove the clamp halves. Lift the handlebars up and back far enough to clear the upper fork yoke, positioning them so that there is no risk of the hydraulic fluid in the reservoir spilling out. Slacken the upper fork yoke clamp bolts and the steering head clamp bolt, and remove the chromium plated steering head cap bolt. The upper yoke can now be knocked upwards, using a hide mallet, and lifted away from the forks.
4 Arrange any remaining cables, leads, etc, so that they do not foul the lower yoke. Obtain a small box or tin in which the steering head balls can be placed safely. Note that as the lower yoke and stem are released, the lower race balls will drop free, and some provision must be made to catch them. As a precaution place a large piece of rag below the headstock to catch any displaced balls. Slacken the lockring, using a C spanner, and lower the steering head stem and lower yoke assembly clear of the frame. Note that there are 20 balls in the lower race and 19 in the upper one, all being of the same size, namely $\frac{1}{4}$ in diameter.
5 Before reassembly, examine and clean the bearing races, then stick the 20 lower race balls into position using high melting point grease. Reassemble the fork unit, following the dismantling sequence in reverse. Note that the steering head should be tightened **just** sufficiently to remove free play. On no account overtighten the head races. It is surprisingly easy to inadvertently apply a loading of several tons to the head bearings, which will quickly break up as a result. When set correctly, there should be no discernible play in the forks when they are shaken. The fork assembly should, however, move easily and without any sign of resistance from lock to lock.

4 Steering head bearings: examination and renovation

1 Before commencing reassembly of the forks, examine the steering head races. The ball bearing tracks of the respective cup and cone bearings should be polished and free from indentations, cracks or pitting. If signs of wear are evident, the cups and cones must be renewed. In order for the straight line steering on any motorcycle to be consistently good, the steering head bearings must be absolutely perfect. Even the smallest amount of wear on the cups and cones may cause steering wobble at high speeds and judder during heavy front wheel braking. The cups and cones are an interference fit on their respective seatings and can be tapped from position with a suitable drift.
2 Ball bearings are relatively cheap. If the originals are marked or discoloured they **must** be renewed. To hold the steel balls in place during reassembly of the fork yokes, pack the bearings with grease. The upper and lower races contain 19 and 20 $\frac{1}{4}$ in steel balls respectively. Although a small gap will remain when the balls have been fitted, on no account must an extra ball be inserted, as the gap is intended to prevent the balls from skidding against each other and wearing quickly.

2.2a Release the speedometer drive cable at the gearbox

2.2b Remove the spindle clamps and remove the front wheel

2.3 Release the mudguard mounting bolts

2.4a Slacken the lower yoke pinch bolt ...

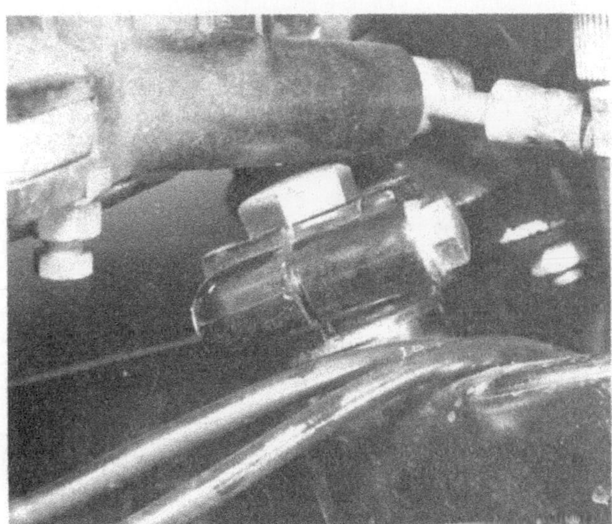
2.4b ... and upper yoke pinch bolt ...

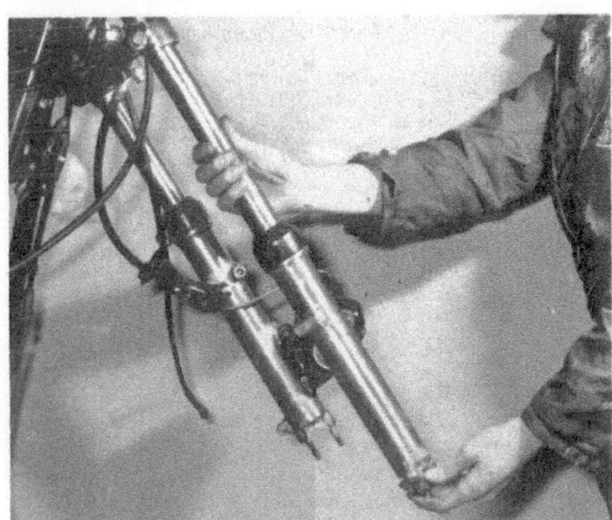
2.5 ... then pull fork leg clear of yokes

3.3 A: Chrome top nut. B: Pinch bolt. C: Adjusting ring

5 Front fork legs: dismantling, renovation and reassembly

1 Having removed the fork legs as described in Section 2, they may be dismantled for further examination. Always deal with one leg at a time,and on no account interchange components from one leg to the other as the various moving parts will have bedded in during use, and should remain matched. Commence by draining the oil, either by way of the drain plug in the lower leg or by removing the bolt and inverting the leg. Pumping the unit will assist in the draining operation.
2 Refer to the accompanying line drawing, then start dismantling, laying each part out on a clean surface as it is removed. Remove the chromium-plated top bolt and withdraw the spacer and spring seat, followed by the fork spring. It will be necessary to prevent the damper assembly from turning while the allen screw in the base of the lower leg is slackened. This will be particularly necessary if the forks are being stripped for the first time since their initial assembly. In the absence of the Kawasaki holding tool (part number 57001 – 142) a little ingenuity must be applied. It was found that a piece of one inch dowel with the end ground to a taper could be used as an improvised holding tool (see photograph), but it may be found ineffective if an excessive amount of thread locking fluid has been used.
3 Having managed to remove the retaining bolt, the stanchion may be pulled out of the lower leg. Shake the damper rod out of the stanchion, then remove the circlip in the lower end of the stanchion and shake out the remaining damper components. The piston may be removed from the damper rod after releasing the circlip which retains it.
4 The parts most liable to wear over an extended period of service are the internal surfaces of the lower leg and the outer surfaces of the fork stanchion or tube. If there is excessive play between these two parts they must be replaced as a complete unit. Check the fork tube for scoring over the length which enters the oil seal. Bad scoring here will damage the oil seal and lead to fluid leakage.
5 It is advisable to renew the oil seals when the forks are dismantled even if they appear to be in good condition. This will save a strip-down of the forks at a later date if oil leakage occurs. The oil seal in the top of each lower fork leg is retained by an internal C ring which can be prised out of position with a small screwdriver. Check that the dust excluder rubbers are not split or worn where they bear on the fork tube. A worn excluder will allow the ingress of dust and water which will damage the oil seal and eventually cause wear of the fork tube.
6 It is not generally possible to straighten forks which have been badly damaged in an accident, particularly when the correct jigs are not available. It is always best to err on the side of safety and fit new ones, especially since there is no easy means to detect whether the forks have been over stressed or metal fatigued. Fork stanchions (tubes) can be checked, after removal from the lower legs, by rolling them on a dead flat surface. Any misalignment will be immediately obvious.
7 The fork springs will take a permanent set after considerable usage and will need renewal if the fork action becomes spongy. The service limit for the total free length of each spring is 456·7 mm (17·9 in). Always renew them as a matched pair.
8 Fork damping is governed by the viscosity of the oil in the fork legs, normally SAE 10W/20, and by the action of the damper assembly. Each fork leg holds 186 – 194 cc of damping fluid.

6 Steering head lock: maintenance

1 A security lock is mounted on the headstock, enabling the owner to immobilise the machine by locking the steering in one position. The lock consists of a key operated plunger which engages in a slot in the steering head. A small return spring disengages the lock mechanism when the key is released.
2 Maintenance is confined to keeping the lock lightly lubricated, using light machine oil or one of the multipurpose aerosol lubricants. In the event that the lock malfunctions, it will be necessary to remove the body after unscrewing the cover plate, and to fit a replacement unit.

7 Frame: examination and renovation

1 The frame is unlikely to require attention unless accident damage has occurred. In some cases, renewal of the frame is the only satisfactory remedy if the frame is badly out of alignment. Only a few frame specialists have the jigs and mandrels necessary for resetting the frame to the required standard of accuracy, and even then there is no easy means of assessing to what extent the frame may have been overstressed.
2 After the machine has covered a considerable mileage, it is advisable to examine the frame closely for signs of cracking or splitting at the welded joints. Rust corrosion can also cause weakness at these joints. Minor damage can be repaired by welding or brazing, depending on the extent and nature of the damage.
3 Remember that a frame which is out of alignment will cause handling problems and may even promote 'speed wobbles'. If misalignment is suspected, as a result of an accident, it will be necessary to strip the machine completely so that the frame can be checked, and if necessary, renewed.

5.2a Remove fork top bolt and spring seat, ...

5.2b ... then withdraw the fork spring

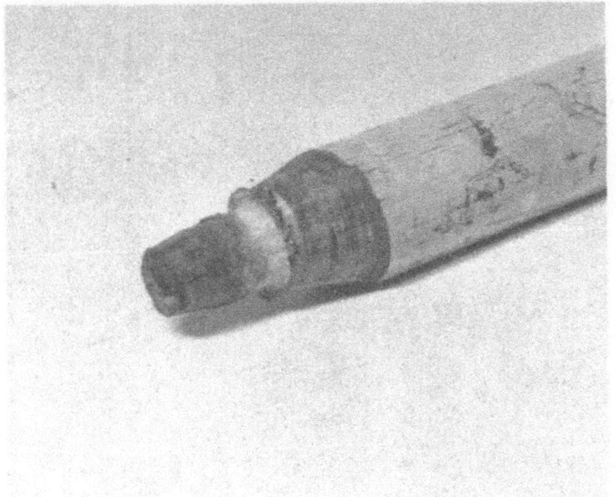

5.2c Tapered dowel can be used to lock damper assembly ...

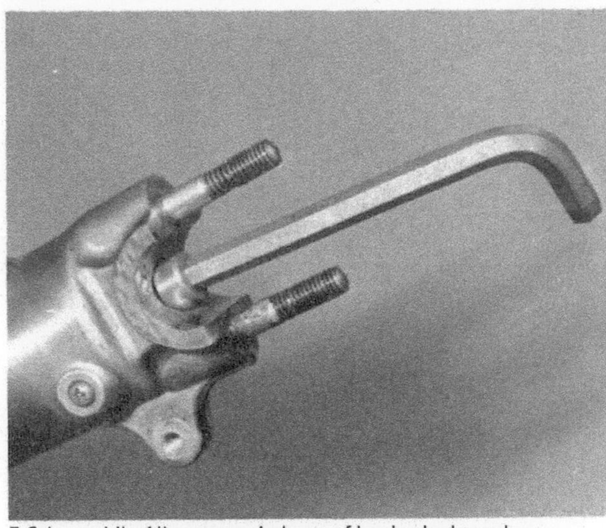

5.2d ... while Allen screw in base of leg is slackened

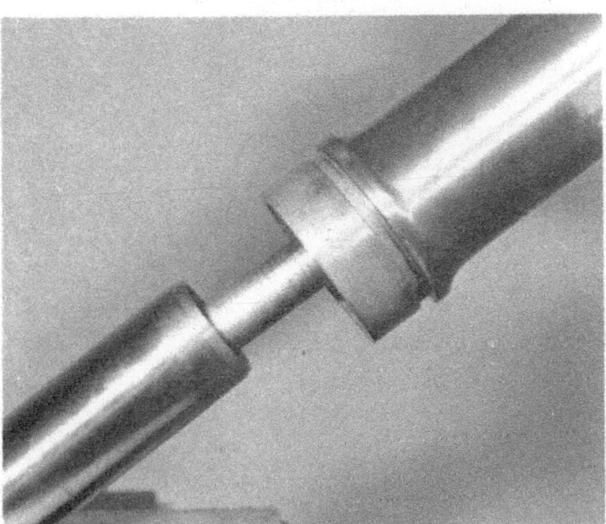

5.3a Pull stanchion and damper out of lower leg

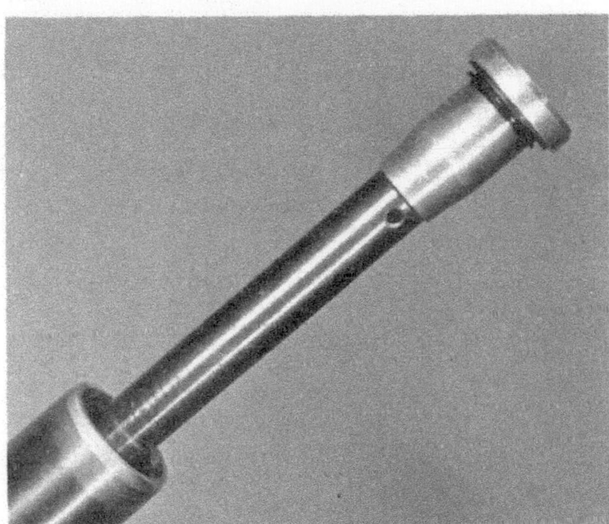

5.3b Alloy seat will lift off the end of damper rod

5.3c Damper assembly is retained by a circlip

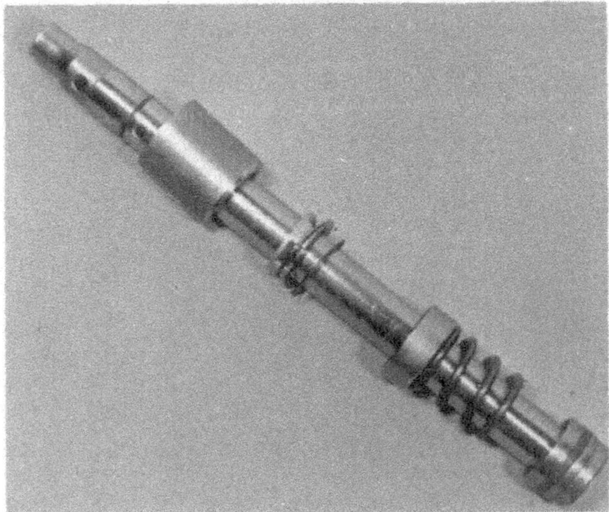

5.4 Examine the moving parts for signs of wear

Fig. 4.1 Front forks

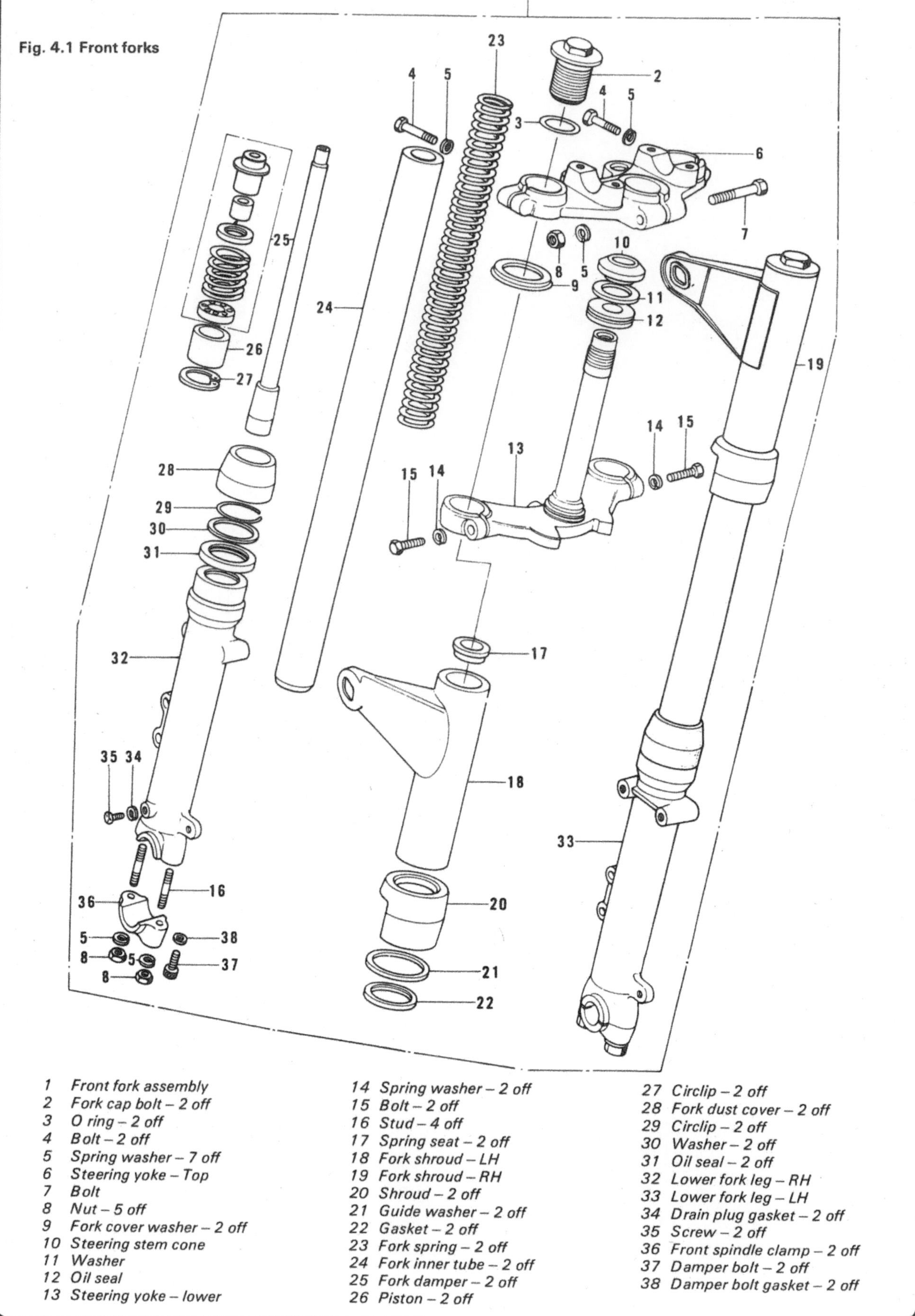

1 Front fork assembly
2 Fork cap bolt – 2 off
3 O ring – 2 off
4 Bolt – 2 off
5 Spring washer – 7 off
6 Steering yoke – Top
7 Bolt
8 Nut – 5 off
9 Fork cover washer – 2 off
10 Steering stem cone
11 Washer
12 Oil seal
13 Steering yoke – lower
14 Spring washer – 2 off
15 Bolt – 2 off
16 Stud – 4 off
17 Spring seat – 2 off
18 Fork shroud – LH
19 Fork shroud – RH
20 Shroud – 2 off
21 Guide washer – 2 off
22 Gasket – 2 off
23 Fork spring – 2 off
24 Fork inner tube – 2 off
25 Fork damper – 2 off
26 Piston – 2 off
27 Circlip – 2 off
28 Fork dust cover – 2 off
29 Circlip – 2 off
30 Washer – 2 off
31 Oil seal – 2 off
32 Lower fork leg – RH
33 Lower fork leg – LH
34 Drain plug gasket – 2 off
35 Screw – 2 off
36 Front spindle clamp – 2 off
37 Damper bolt – 2 off
38 Damper bolt gasket – 2 off

5.5a Prise out and renew the oil seals, if worn or damaged

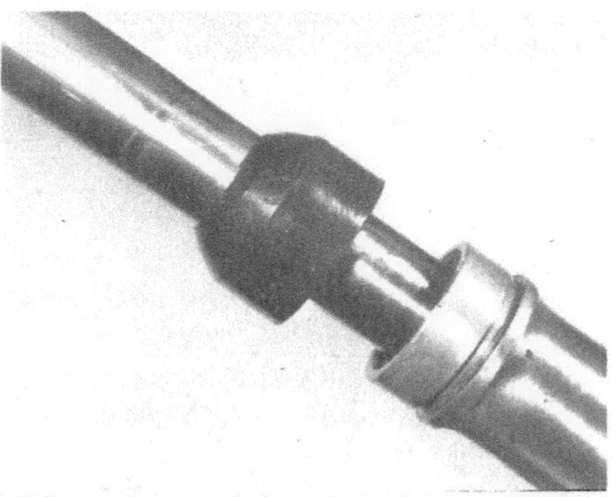
5.5b Dust excluder must be in good condition to prevent sealwear

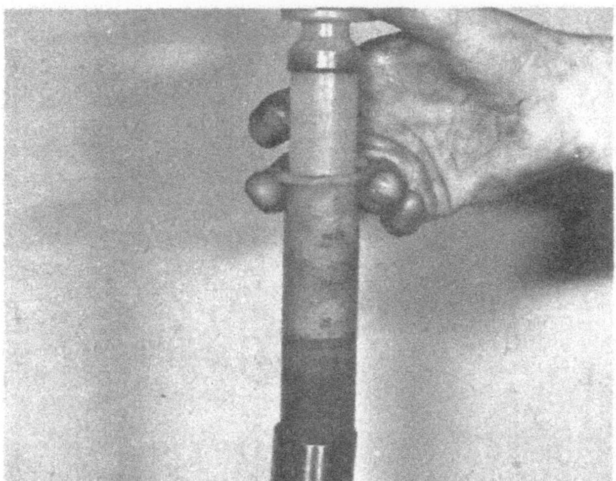
5.8 Refill fork legs with the correct quantity of oil

6.1 Steering lock is incorporated in headstock

8 Swinging arm fork: removal and renovation

1 The swinging arm fork is supported on two headed bushes which pivot on an inner sleeve. The assembly is retained by a long pivot shaft which passes through lugs on the frame and through the centre of the sleeve. A grease nipple is fitted to enable grease to be pumped to the bearing surfaces.
2 Wear in the swinging arm bushes is characterised by a tendency for the rear of the machine to twitch when ridden hard through a series of bends. This can be checked by placing the machine on the centre stand, and pushing the swinging arm from side to side. Any discernible free play will necessitate the removal of the swinging arm for further examination.
3 Commence by detaching the silencer from each side of the machine, after slackening the clamp bolt and mounting nuts on each unit. Detach the rear brake switch operating spring, then remove the brake adjusting nut. The brake operating rod can now be disengaged from the operating lever. Pull out the spring pin from the torque arm mounting stud, then remove the securing nut and disengage the torque arm. Release the chainguard mounting nuts, and lift the guard away.
4 Remove the split pin which retains the rear wheel spindle nut, then slacken the nut. Release and back off the chain tensioner drawbolts, and swing them through 90° degrees. The wheel can now be pushed forwards as far as possible, and the final drive chain disengaged from the rear wheel sprocket. Slacken the bolt which retains each of the stops in the end of the forks. Remove the stops, then pull the wheel backwards until it clears the frame. Lift it clear and place it to one side to await reassembly.
5 Remove the lower suspension mounting bolt from each side of the machine, and push the units clear of the swinging arm. Slacken the pivot shaft nut and pull the pivot shaft out, supporting the swinging arm. The swinging arm can now be drawn rearwards, noting that the caps on each end of the pivot tube will probably drop clear. Disengage the fork from the drive chain, and place it on a bench to await further dismantling.
6 Displace the pivot sleeve and wash the sleeve and the headed bushes to remove all trace of grease. Examine the sleeve and bush bearing surfaces for signs of wear or scoring. If damaged or worn, or if below the limits given in the Specifications Section, replace the components as necessary. The two headed bushes may be driven out using a suitable bar or drift. When fitting new bushes, ensure that they are tapped squarely into the swinging arm bore, and that they seat securely. Clean off any corrosion on the inner sleeve if this is to be re-used.
7 If, on reassembly, there appears to be excessive axial play in the swinging arm, it is possible to correct this by placing appropriate thicknesses of shims under the end caps. When the unit has been installed, use a grease gun to force grease into the assembly. Continue greasing until it exudes at each end, wiping off any excess.

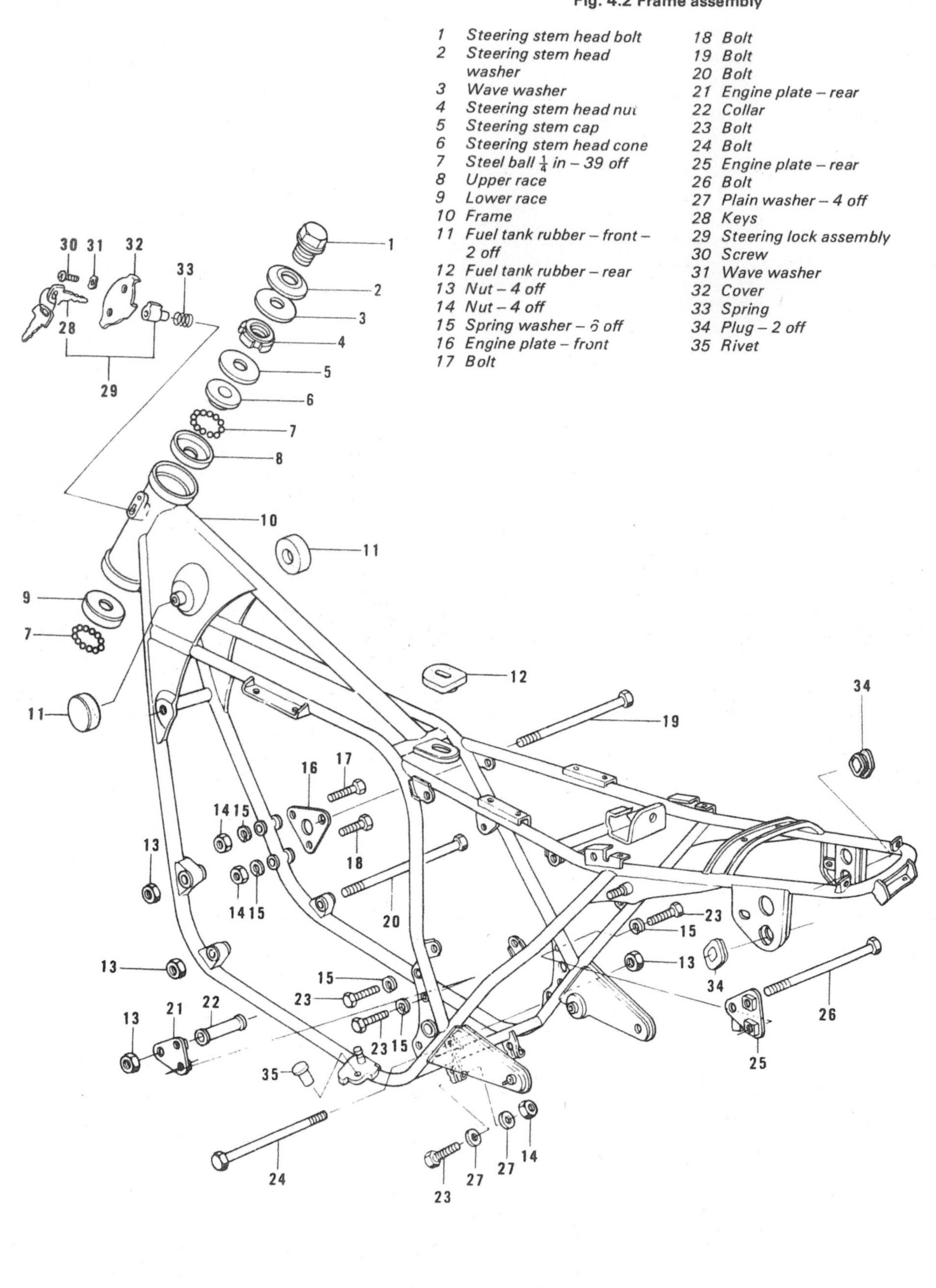

Fig. 4.2 Frame assembly

1 Steering stem head bolt
2 Steering stem head washer
3 Wave washer
4 Steering stem head nut
5 Steering stem cap
6 Steering stem head cone
7 Steel ball $\frac{1}{4}$ in – 39 off
8 Upper race
9 Lower race
10 Frame
11 Fuel tank rubber – front – 2 off
12 Fuel tank rubber – rear
13 Nut – 4 off
14 Nut – 4 off
15 Spring washer – 6 off
16 Engine plate – front
17 Bolt
18 Bolt
19 Bolt
20 Bolt
21 Engine plate – rear
22 Collar
23 Bolt
24 Bolt
25 Engine plate – rear
26 Bolt
27 Plain washer – 4 off
28 Keys
29 Steering lock assembly
30 Screw
31 Wave washer
32 Cover
33 Spring
34 Plug – 2 off
35 Rivet

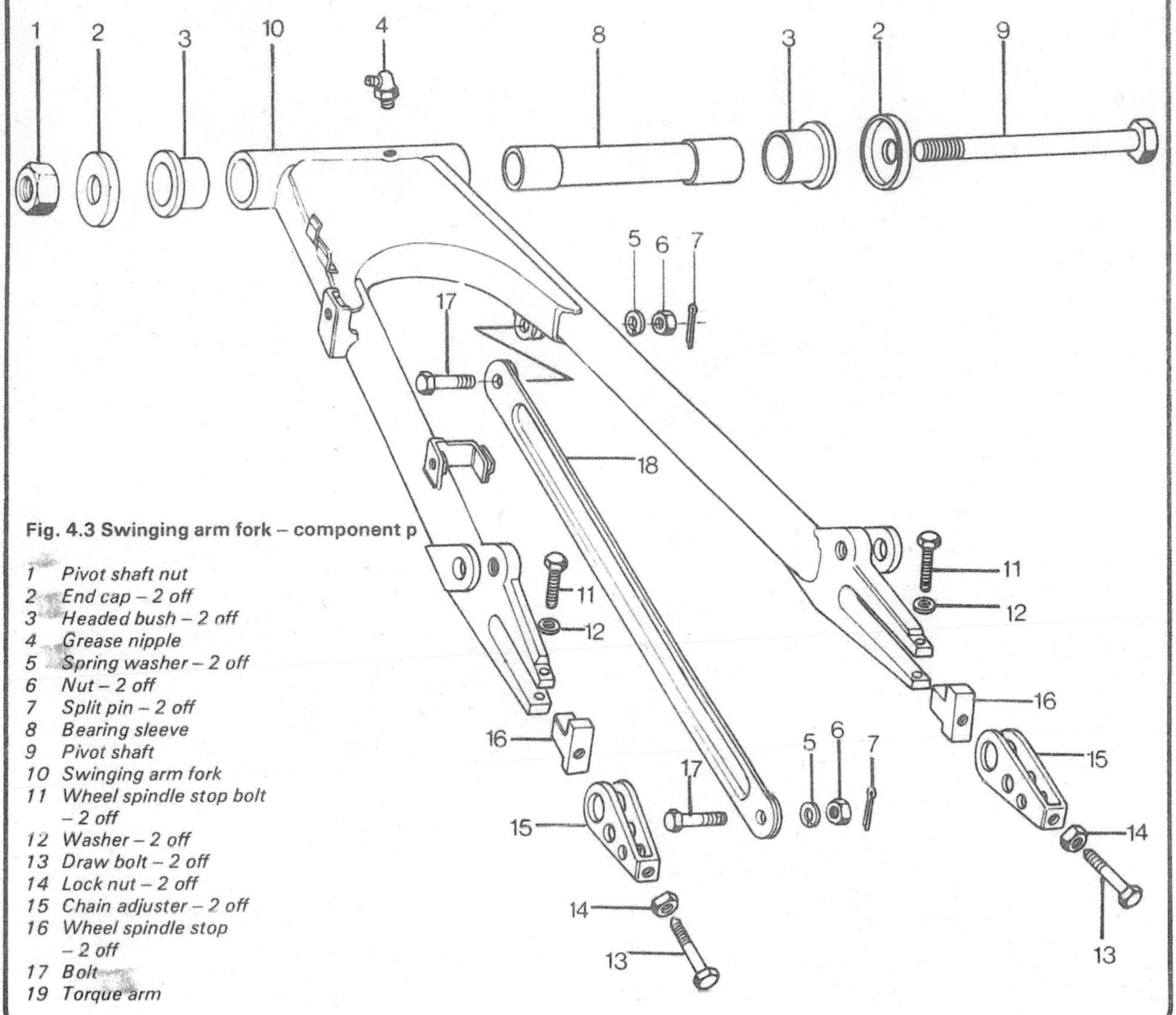

Fig. 4.3 Swinging arm fork – component p

1 *Pivot shaft nut*
2 *End cap – 2 off*
3 *Headed bush – 2 off*
4 *Grease nipple*
5 *Spring washer – 2 off*
6 *Nut – 2 off*
7 *Split pin – 2 off*
8 *Bearing sleeve*
9 *Pivot shaft*
10 *Swinging arm fork*
11 *Wheel spindle stop bolt – 2 off*
12 *Washer – 2 off*
13 *Draw bolt – 2 off*
14 *Lock nut – 2 off*
15 *Chain adjuster – 2 off*
16 *Wheel spindle stop – 2 off*
17 *Bolt*
19 *Torque arm*

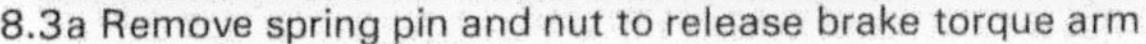

8.3a Remove spring pin and nut to release brake torque arm

8.3b Detach chain guard

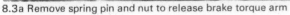

8.4 Release chain, wheel spindle nut and drawbolt assembly

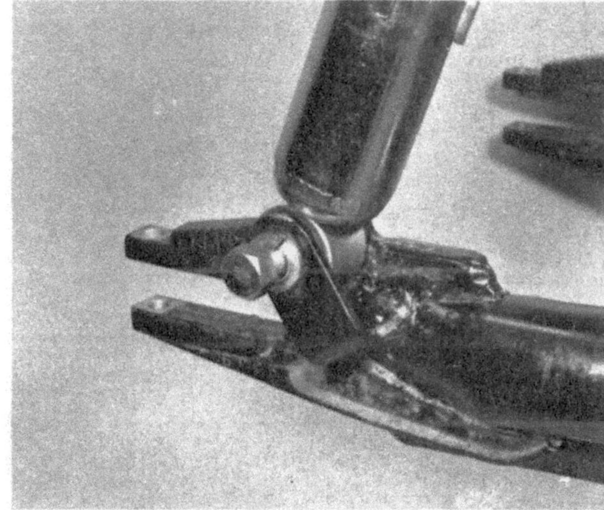
8.5a Remove the rear wheel, and detach suspension units

8.5b Slacken and remove the pivot shaft nut, ...

8.5c ... withdraw the pivot shaft ...

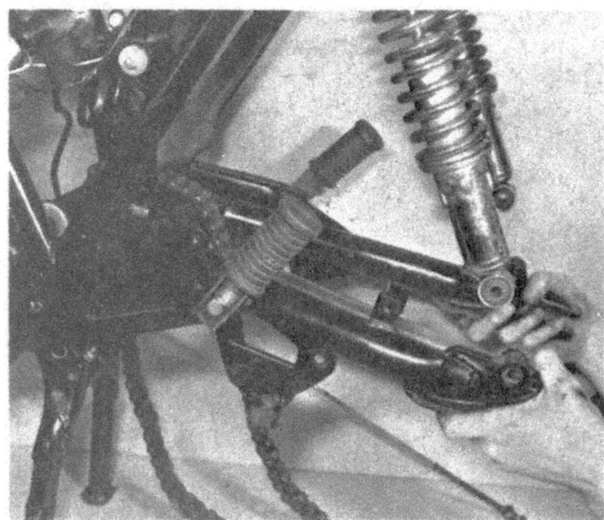
8.5d ... and disengage the swinging arm

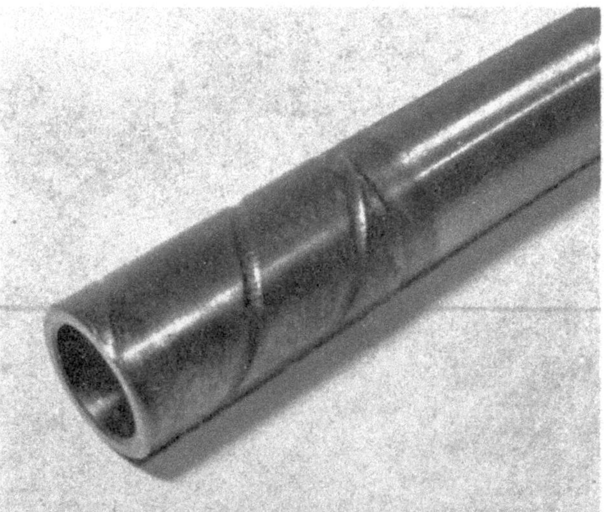
8.6 Clean and examine the inner sleeve and bushes

8.7a Grease all components during reassembly

8.7b Do not forget to refit end caps prior to installation

8.7c Bushes should be kept well lubricated via the grease nipple

9 Rear suspension units: examination

1 Rear suspension units of the 3 way adjustable type with hydraulic damping are fitted to the Z650. The units can be adjusted to give 3 different settings. A hook spanner in the toolkit is used to adjust the units by means of peg holes.
2 There is no means of draining or topping up the units as they are permanently sealed. In the interests of good road holding, both units should be renewed if either starts to leak or loses its damping action.

10 Centre stand: examination

1 The centre stand is attached to the machine by two bolts on the bottom of the frame. It is returned by a centre spring. The bolts and spring should be checked for tightness and tension respectively. A weak spring can cause the centre stand to ground on corners and unseat the rider.

11 Prop stand: examination

1 The prop stand is secured to a plate on the frame with a bolt and nut, and is retracted by a tension spring. Make sure the bolt is tight and the spring is not overstretched, otherwise an accident can occur if the stand drops during cornering.

12 Footrests and rear brake pedal: examination

1 The footrests are of the swivel type and are retained by a clevis pin secured by a split pin. The advantage of this type of footrest is that if the machine should fall over the footrest will fold up instead of bending.
2 The rear brake pedal is held in position by a stud and domed nut, the pedal return spring must be detached to remove the brake lever.

13 Dualseat: removal and replacement

1 The dualseat is attached to the frame by two clevis pins that are located with split pins, on the right-hand side of the frame. To remove the seat, release the spring loaded catch on

9.1 Rear suspension units can be adjusted for preload as shown

the left-hand side, and prop the seat up with the stay provided. Withdraw the two split pins from the clevis pivot pins, and remove the pivot pins. The seat mountings and damper rubbers can be left in place as the seat is lifted off.

2 If the dualseat is removed because it is torn, it is possible in most cases to find a specialist firm that recovers dualseats for an economical price, usually considerably cheaper than having to buy a new replacement. The usual charge is about 50% the cost of a new replacement, depending on the extent of the damage.

14.1 Instrument cowls are each retained by a single screw. Note trip counter reset on side

14 Speedometer and tachometer heads: removal and replacement

1 The speedometer and tachometer are both mounted together on a single panel on top of the front forks. They are mounted on studs with rubber bushes and secured with nuts. The heads are encased in light alloy shrouds secured to the instruments by a single crosshead screw. The shrouds have to be removed first, to enable the instruments to be released.

2 After the shrouds are detached the drive cables can be unscrewed. The rubber mounted bulb holders can be pulled out with the bulbs. Check for blown bulbs while they are out. The four bulbs in the dash panel are also a push fit and can be checked at the same time.

3 The speedometer and tachometer heads cannot be repaired by the private owner, and if a defect occurs a new instrument has to be fitted. Remember that a speedometer in correct working order is required by law on a machine in the UK also many other countries.

4 Speedometer and tachometer cables are only supplied as a complete assembly. Make sure the cables are routed correctly through the clamps provided on the top fork yoke, brake branch pipe, and the frame.

15 Speedometer and tachometer drives: location and examination

1 The speedometer is driven from a gear inside the front wheel hub assembly. The gear is driven internally by a tongued washer (receiver). The receiver engages with two slots in the wheel hub, on the left-hand side. As the whole gearbox is pre-packed with grease on assembly, it should last the life of the machine, or until new parts are fitted. The spiral pinion that drives off the internal gear is retained in the speedometer gearbox casing by a grub screw, which should always be secured tightly.

2 The tachometer drive runs off the camshaft in the cambox and screws directly into the cylinder head cover in the centre position. The cable is retained by a screwed ferrule, in the same manner as the speedometer cable.

16 Cleaning the machine

1 After removing all the surface dirt with warm water and a rag or sponge, use a cleaning compound such as 'Gunk' or 'Jizer' for the oily parts. Apply the cleaner with a brush when the parts are clogged so that it has an opportunity to soak into the film of oil or grease.

Finish off by washing down liberally, taking care that water does not enter into the carburettors, air cleaner or electrics. If desired, a polish such as Solvol Autosol can be applied to the alloy parts to give them a full lustre. Application of a wax polish to the cycle parts and a good chrome cleaner to the chrome parts will also give a good finish. Always wipe down the machine if used in the wet, and make sure the chain is well oiled. Check that the control cables are kept well oiled (this will only take 5 minutes of your time each week with an oil can). There is also less chance of water getting into the cables, if they are well lubricated.

17 Fault diagnosis: frame and forks

Symptom	Cause	Remedy
Machine veers to left or right with hands off handlebars	Wheels out of alignment Forks twisted Frame bent	Check wheels and realign. Strip and repair. Strip and repair or renew.
Machine tends to roll at low speeds	Steering head bearings not adjusted correctly or worn	Check adjustment and renew the bearings, if worn.
Machine tends to wander	Worn swinging arm bearings	Check and renew bearings. Check adjustment and renew.
Forks judder when front brake is applied	Steering head bearings slack Forks worn on sliding surfaces	Strip forks. Check adjustment, renew all worn parts.
Forks bottom	Short of oil	Replenish with correct viscosity oil.
Fork action stiff	Fork legs out of alignment Bent shafts, or twisted yokes	Strip and renew or slacken clamp bolts, front wheel spindle and top bolts. Pump forks several times, and tighten from bottom upwards.
Machine tends to pitch badly	Defective rear suspension units, or ineffective fork damping	Check damping action. Check the grade and quantity of oil in the front forks.

Chapter 5 Wheels, brakes and tyres

Contents

Specifications

Tyres

Front	3·25H – 19 in 4PR	
Rear	4·00H – 18 in 4PR	

Tyre pressures

	Solo	Pillion or high speed
Front	2·0 kg/cm² (28 psi)	2·0 kg/cm² (28 psi)
Rear	2·25 kg/cm² (32 psi)	2·50 kg/cm² (36 psi)

Brakes

Front	245 mm hydraulic disc brake
Rear	180 x 40 mm single leading shoe drum brake

1 General description

The Z650 models have a 19 inch diameter front wheel and an 18 inch diameter rear wheel. The front tyre is of the ribbed tread pattern; the rear tyre has a block tread pattern. All models employ steel rims in conjunction with cast aluminium hubs. The front brake is of the hydraulic disc type, the rear brake is the internal expanding drum type.

2 Front wheel: examination and renovation

1 Place the machine on its centre stand so that the front wheel is clear of the ground. Spin the wheel by hand and check the rim for alignment. Small irregularities can be corrected by tightening the spokes in the affected area. Any flats in the wheel rim will be evident at the same time. In this latter case it will be necessary to have the wheel rebuilt with a new rim. The machine should not be run with a deformed wheel, since this will have a very adverse effect on handling.
2 Check for loose or broken spokes. Tapping the spokes is a good guide to the correct tension; a loose spoke will always produce a different sound and should be tightened by turning the nipple in an anti-clockwise direction. Always check for run out by spinning the wheel again. If the spokes have to be tightened by an excessive amount, it is advisable to remove the tyre and tube as detailed in Section 15 of this Chapter. This will enable the protruding ends of the spokes to be ground off, thus preventing them from chafing the inner tube and causing punctures.

3 Front wheel disc brake: examination and renovation

1 Check the front brake master cylinder, hose and caliper unit for signs of fluid leakage. Pay particular attention to the condition of the synthetic rubber hose, which should be renewed without question if there are signs of cracking, splitting or other exterior damage.
2 Check the level of hydraulic fluid by removing the cap on the brake fluid reservoir and lifting out the diaphragm and diaphragm plate. This is one of the maintenance tasks which should never be neglected. Make certain that the handlebars are in the central position when removing the reservoir cap, because if the fluid level is high, the fluid will spill over the reservoir brim. If the level is particularly low, the fluid delivery

passage will be allowed direct contact with the air and may necessitate the bleeding of the system at a later date. A level mark is given on the inside of the reservoir cylinder; if the level is below the mark, brake fluid of the correct grade must be added. **NEVER USE ENGINE OIL** or anything other than the recommended fluid. Other fluids have unsatisfactory characteristics and will quickly destroy the seals.

3 The brake pads should be inspected for wear. Each pad is marked with a scribed line, indicating the maximum wear limit. If worn beyond this point, **both** brake pads must be renewed as a set. The brake pads can be checked while they are still in position in the caliper and the front wheel is still in situ. If the front brake is operated, the extent of wear can be easily seen.

4 The brake pads can be removed from the caliper after the front wheel has been taken out. Commence by removing the retaining screw and backplate on the inside pad. The pad will push out of position. **Very gently** apply the front brake lever, which will operate the caliper piston and so push the outer pad from position. Do not pump the brake when carrying out this operation or there is a danger of the piston being pushed out of the cylinder. It will be noted that the outer pad (piston pad) has a steel shim on the rear face. The shim is fitted to prevent the disc brake assembly squeaking during operation, and is located by a small projection on the brake pad.

When fitting new pads, it will probably be found that the increased size of the pads will prevent the brake disc from fitting between the pads when the front wheel is being replaced. To overcome this, press hard on the outer (piston) pad and at the same time slightly loosen the brake bleed valve. The pad will move inwards slowly and then stop, at which point the bleed valve must be tightened immediately. It will be found that a small amount of fluid will have been ejected from the bleed valve. Wipe up the fluid immediately and then check the level in the master cylinder.

5 If brake action becomes spongy, or if any part of the hydraulic system is dismantled (such as when the hose is replaced) it is necessary to bleed the system in order to remove all traces of air. The following procedure should be followed:

6 Attach a tube to the bleed valve at the top of the caliper unit, after removing the dust cap. It is preferable to use a transparent plastic tube, so that the presence of air bubbles is seen more readily.

7 The far end of the tube should rest in a small bottle so that it is submerged in hydraulic fluid. This is essential, to prevent air from passing back into the system. In consequence, the end of the tube must remain submerged at all times.

8 Check that the reservoir on the handlebars is full of fluid and replace the cap to keep the fluid clean.

9 If spongy brake action necessitates the bleeding operation, squeeze and release the brake lever several times in rapid succession, to allow the pressure in the system to build up. Then open the bleed valve by unscrewing it one complete turn whilst maintaining pressure on the lever. This is a two-person operation. Squeeze the lever fully until it meets the handlebar, then close the bleed valve. If parts of the system have been replaced, the bleed valve can be opened from the beginning and the brake lever worked until fluid issues from the bleed tube. Note that it may be necessary to top up the reservoir during this operation; if it empties, air will enter the system and the whole operation will have to be repeated.

10 Repeat operation 9 until bubbles disappear from the bleed tube. Close the bleed valve fully, remove the bleed tube and replace the dust cap.

11 Check the level in the reservoir and top up if necessary. Never use the fluid which has drained into the bottle at the end of the bleed tube because this contains air bubbles which will re-introduce air into the system. The fluid must stand for 24 hours before it can be re-used.

12 Refit the diaphragm and diaphragm plate and tighten the reservoir cap securely.

13 Do not spill fluid on the cycle parts. It is a very effective paint stripper! Also, the plastic glasses in the speedometer and tachometer heads will be badly obscured if fluid is spilt on them.

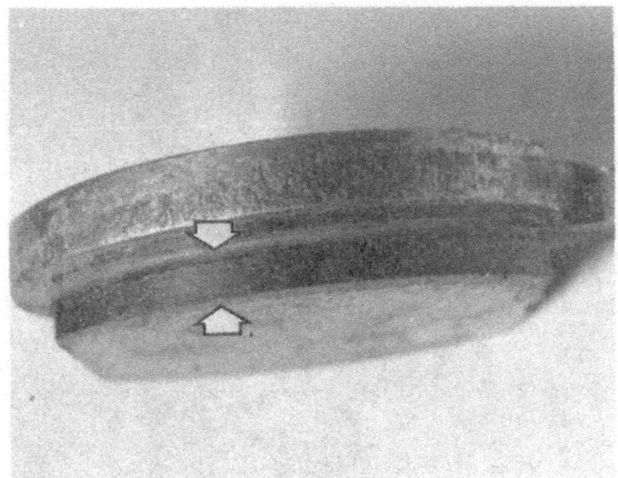

3.3 Maximum wear limit is denoted by scribed line

3.4a Release the retaining screw and backplate

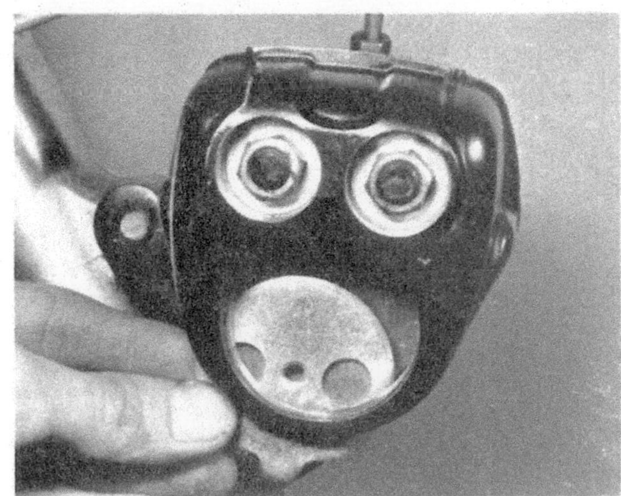

3.4b Fixed pad can now be displaced and withdrawn, ...

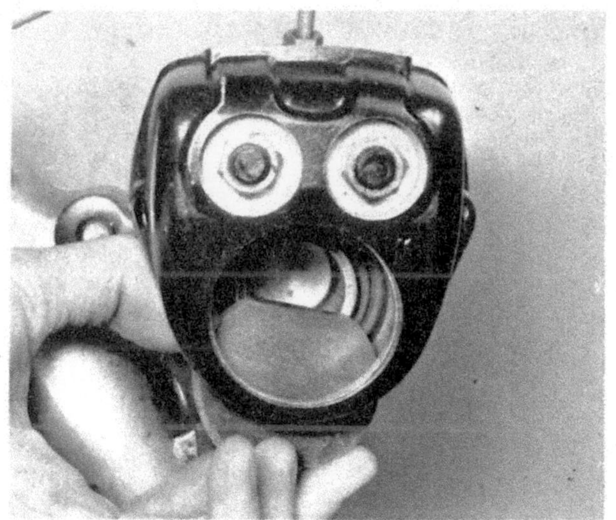

3.4c ... followed by inner, or moving, pad

3.6 Bleed tube must be in position before valve is slackened

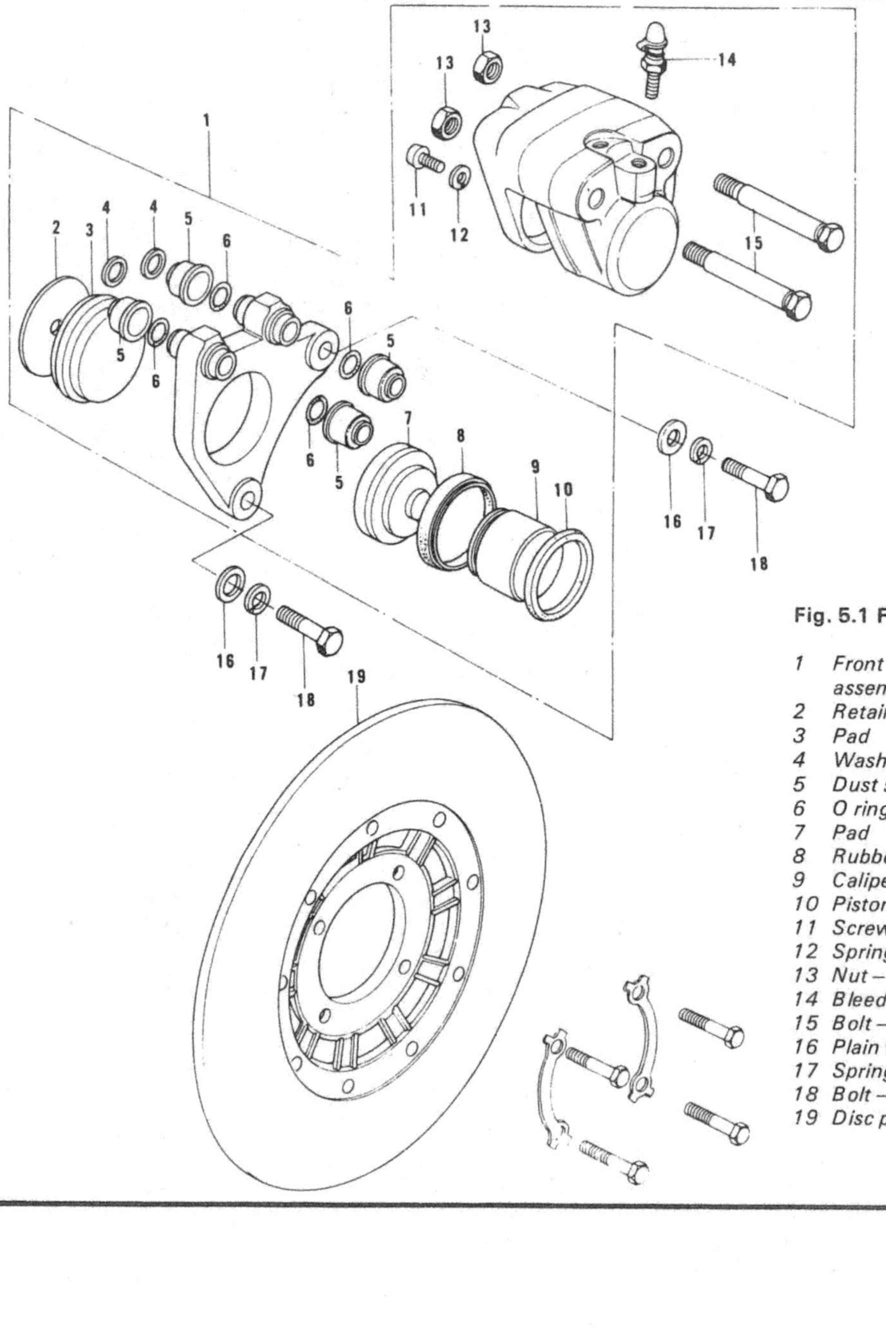

Fig. 5.1 Front brake caliper assembly

1 *Front brake caliper assembly*
2 *Retaining plate*
3 *Pad*
4 *Washer – 2 off*
5 *Dust seal – 4 off*
6 *O ring – 4 off*
7 *Pad*
8 *Rubber dust seal*
9 *Caliper piston*
10 *Piston seal*
11 *Screw*
12 *Spring washer*
13 *Nut – 2 off*
14 *Bleed screw*
15 *Bolt – 2 off*
16 *Plain washer – 2 off*
17 *Spring washer – 2 off*
18 *Bolt – 2 off*
19 *Disc plate*

4 Removing and replacing the brake disc

1 It is unlikely that the disc will require attention until a considerable mileage has been covered, unless premature scoring of the disc has taken place thereby reducing braking efficiency. To remove the disc, first detach the front wheel as described in Chapter 4, Section 2.1 and 2. The disc is bolted to the front wheel on the left-hand side by four bolts, which are secured in pairs by a common tab washer. Bend back the tab washers and remove the bolts, to free the disc.

2 The brake disc can be checked for wear and for warpage whilst the front wheel is still in the machine. Using a micrometer measure the thickness of the disc at the point of greatest wear. If the measurement is much less than the recommended service limit of 6 mm (0·236 in) the disc should be renewed. Check the warpage of the disc by setting up a suitable pointer close to the outer periphery of the disc and spinning the front wheel slowly. If the total warpage is more than 0.3mm (0.012 in) the disc should be renewed. A warped disc, apart from reducing the braking efficiency, is likely to cause juddering during braking and will also cause the brake to bind when it is not in use.

5 Master cylinder: examination and renovation

1 The master cylinder is unlikely to give trouble unless the machine has been stored for a lengthy period or until a considerable mileage has been covered. The usual signs of trouble are leakage of hydraulic fluid and a gradual fall in the fluid reservoir content.

2 To gain full access to the master cylinder, commence the dismantling operation by attaching a bleed tube to the caliper unit bleed nipple. Open the bleed nipple one complete turn, then operate the front brake lever until all fluid is pumped out of the reservoir. Close the bleed nipple, detach the tube and store the fluid in a closed container for subsequent re-use.

3 Detach the hose and also the stop lamp switch. Remove the handlebar lever pivot bolt and the lever itself.

4 Access is now available to the piston and the cylinder and it is possible to remove the piston assembly, together with all the relevant seals. Take note of the way in which the seals are arranged because they must be replaced in the same order. Failure to observe this necessity will result in brake failure.

5 Clean the master cylinder and piston with either hydraulic fluid or alcohol. On no account use either abrasives or other solvents such as petrol. If any signs of wear or damage are evident, renewal is necessary. It is not practicable to reclaim either the piston or the cylinder bore.

6 Soak the new seals in hydraulic fluid for about 15 minutes prior to fitting, then reassemble the parts in **IN EXACTLY THE SAME ORDER**, using the reversal of the dismantling procedure. Lubricate with hydraulic fluid and make sure the feather edges of the various seals are not damaged.

7 Refit the assembled master cylinder unit to the handlebar, and reconnect the handlebar lever, hose, stop lamp etc. Refill the reservoir with hydraulic fluid and bleed the entire system by following the procedure detailed in Section 3.5 of this Chapter.

8 Check that the brake is working correctly before taking the machine on the road, to restore pressure and align the pads with the disc surface.

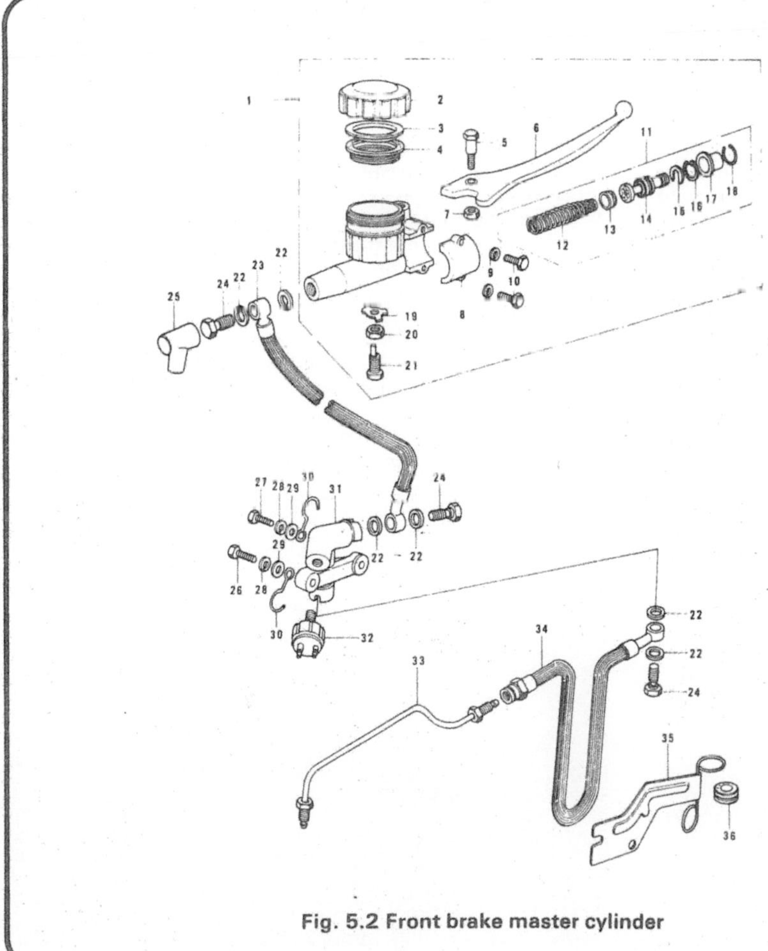

Fig. 5.2 Front brake master cylinder

1 Front brake master cylinder assembly
2 Oil reservoir cap
3 Master cylinder plate
4 Diaphragm
5 Lever bolt
6 Front brake lever
7 Nut
8 Master cylinder holder
9 Plain washer – 2 off
10 Bolt – 2 off
11 Repair kit
12 Return spring
13 Primary cup
14 Piston
15 Stopper
16 Circlip
17 Dust cover
18 Circlip
19 Lock washer
20 Lock nut
21 Lever adjustment screw
22 Oil bolt washer – 6 off
23 Cylinder brake hose
24 Bolt – 3 off
25 Dust cover
26 Bolt
27 Bolt
28 Spring washer – 2 off
29 Washer – 2 off
30 Speedometer and tachometer cable clamp – 2 off
31 Three-way joint
32 Front brake switch
33 Bridge pipe
34 Caliper brake hose
35 Bracket
36 Brake hose grommet

4.1 Brake disc is retained by four bolts and double tab washers

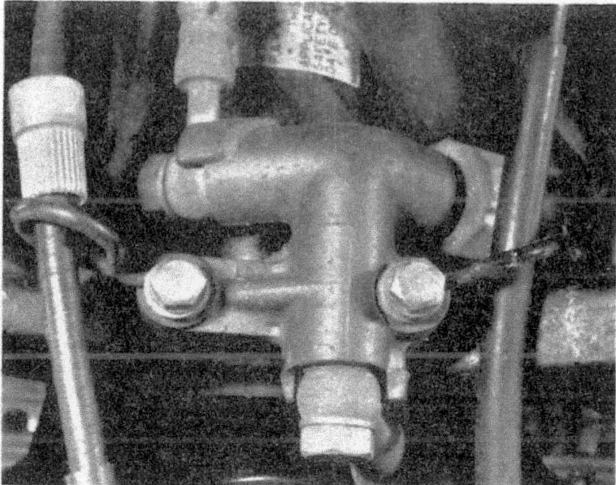
5.1 Examine hydraulic hoses and distributor block for signs of leakage

6 Front wheel bearings: examination and replacement

1 Access is available to the front wheel bearings when the speedometer and front wheel spindle are removed. The bearings are of the ball journal type and non-adjustable. There are two bearings and two oil seals, the two bearings are interposed by a distance collar in the centre of the hub.
2 First remove the speedometer cable by undoing the knurled nut. Remove the front wheel spindle, and pull off the speedometer drive gearbox. Remove the brake disc after releasing the four retaining bolts. Take off the collar and wheel cap, and drive out the left-hand bearing, using a double diameter drift from the right-hand side.

When the bearing is removed, the distance collar can be taken out. Working from inside the hub, use the same drift to displace the right-hand bearing. Remove the oil seal, take out the retaining ring, and from the left-hand side use the drift to tap evenly around the inner race of the right-hand bearing and knock it out.
3 Remove all the old grease from the hub and bearings, wash the bearings in petrol, and dry them thoroughly. Check the bearings for roughness by spinning them whilst holding the inner track with one hand and rotating the outer track with the other. If there is the slightest sign of roughness renew them.
4 Before driving the bearings back into the hub, pack the hub with new grease and also grease the bearings. Use the same double diameter drift to place them into position. Refit any oil seals or dust covers which have been displaced.

7 Front wheel: reassembly and replacement

1 Refit the speedometer gearbox, and fit the wheel cap or shroud which is retained by two bolts.
2 Have the bottom fork clamps ready when the front wheel is lifted back into position. First tighten the front spindle clamp bolt and then the rear bolt for each fork leg, so that there will be a gap at the rear after tightening. Spin the wheel to make sure it revolves freely, and check that the brake operates correctly. Turn the front wheel while inserting the speedometer cable, so that the tongue of the speedometer drive will locate correctly.

8 Rear wheel assembly: examination and renovation

1 Place the machine on the centre stand so that the rear wheel is raised clear of the ground. Check the rim for alignment, damage to the rim or broken spokes by following the procedure relating to the front wheel described in Section 2 of this Chapter.
2 To remove the rear wheel, refer to Chapter 4, Section 8.3 and 4. Note that the right-hand silencer only need be removed.
3 Remove the wheel spindle and take out the brake plate. Take the coupling assembly from the cush drive rubbers, and remove the rubbers. The rear wheel sprocket can be unbolted for inspection, by removing the six nuts and the three locktabs.

6.2a Remove the wheel spindle and speedometer gearbox ...

6.2b ... and remove spacer and hub cover

6.2c Oil seal can be prised out of hub

6.2d Wheel bearing is retained by a circlip

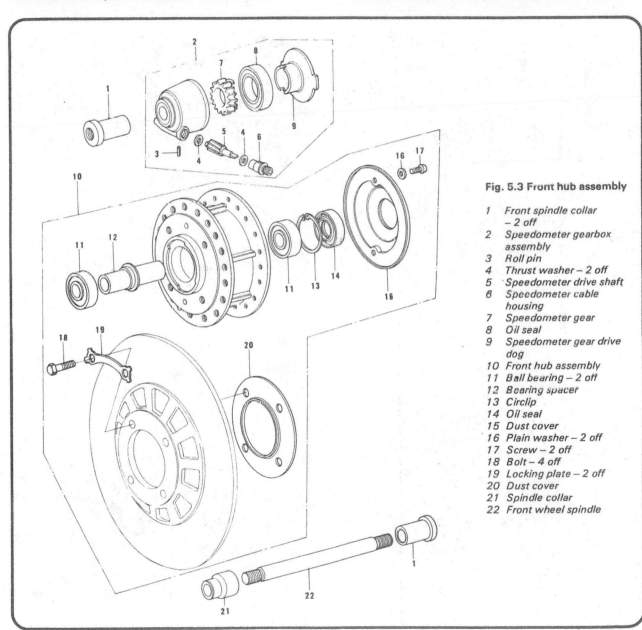

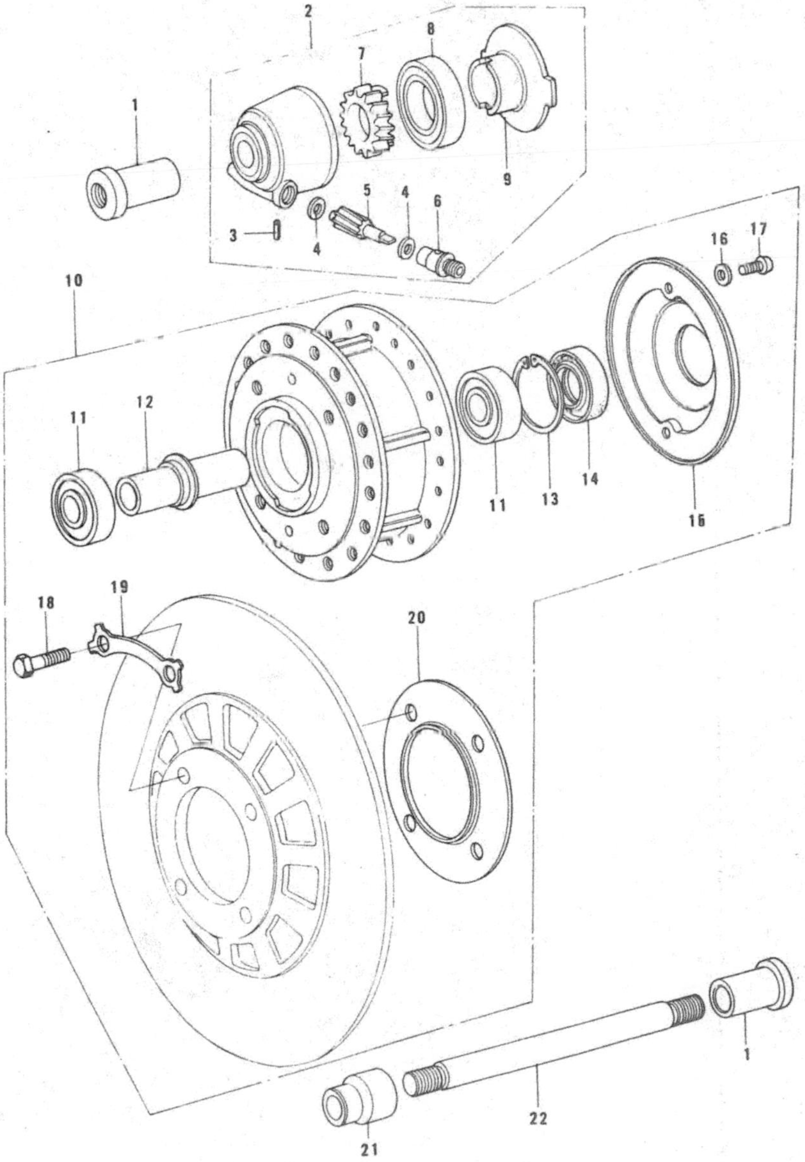

Fig. 5.3 Front hub assembly

1 *Front spindle collar – 2 off*
2 *Speedometer gearbox assembly*
3 *Roll pin*
4 *Thrust washer – 2 off*
5 *Speedometer drive shaft*
6 *Speedometer cable housing*
7 *Speedometer gear*
8 *Oil seal*
9 *Speedometer gear drive dog*
10 *Front hub assembly*
11 *Ball bearing – 2 off*
12 *Bearing spacer*
13 *Circlip*
14 *Oil seal*
15 *Dust cover*
16 *Plain washer – 2 off*
17 *Screw – 2 off*
18 *Bolt – 4 off*
19 *Locking plate – 2 off*
20 *Dust cover*
21 *Spindle collar*
22 *Front wheel spindle*

7.1a Note driving tangs in wheel centre, ...

7.1b ... which engage in speedometer gearbox

7.1c Hub cover is retained by two screws

8.2 Rear wheel should be removed as described in Chapter 4

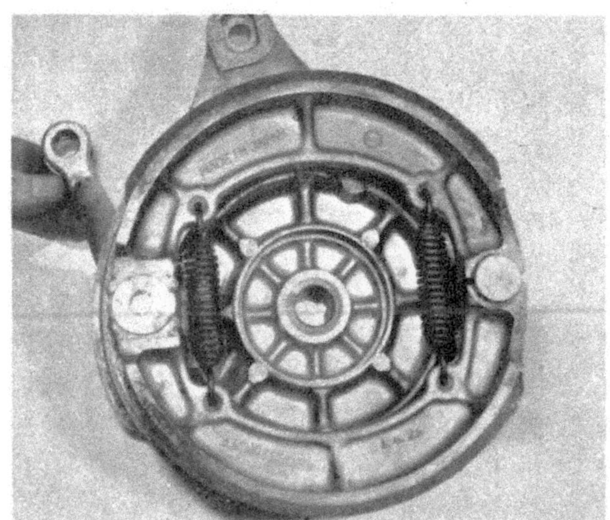
8.3a Remove brake backplate, ...

8.3b ... and cush drive/sprocket unit

9 Rear wheel bearings: examination and replacement

1 The rear wheel bearings are a drive fit into the hub. They are separated by a spacer and a distance collar. There are two bearings in the hub and one in the centre of the rear final drive sprocket boss.

2 Remove the distance collar from the cush drive hub, take out the wheel spindle collar, and then the oil seal. Tapping evenly around the inner race from the inside of the coupling, knock out the bearing.

3 To remove the two bearings from the wheel hub, use a double diameter drift again and tapping evenly around the inner race from the sprocket side knock out the bearing on the brake plate side.

4 Remove the large distance collar, and tapping on the inner race from the brake plate side, knock out the bearing on the sprocket side.

5 Remove all the old grease from the bearings and hub. Wash the bearings in petrol and dry them thoroughly. Check the bearings for roughness by spinning them whilst holding the inner track with one hand, and rotating the outer track with the other hand. If there is the slightest sign of roughness renew them.

6 Before driving the bearings back into the hub and sprocket centre, pack the hub with new grease and also grease the bearings. Use the same double diameter drift to place them into position. Refit any oil seals or dust covers which have been displaced, renewing them if damaged or worn.

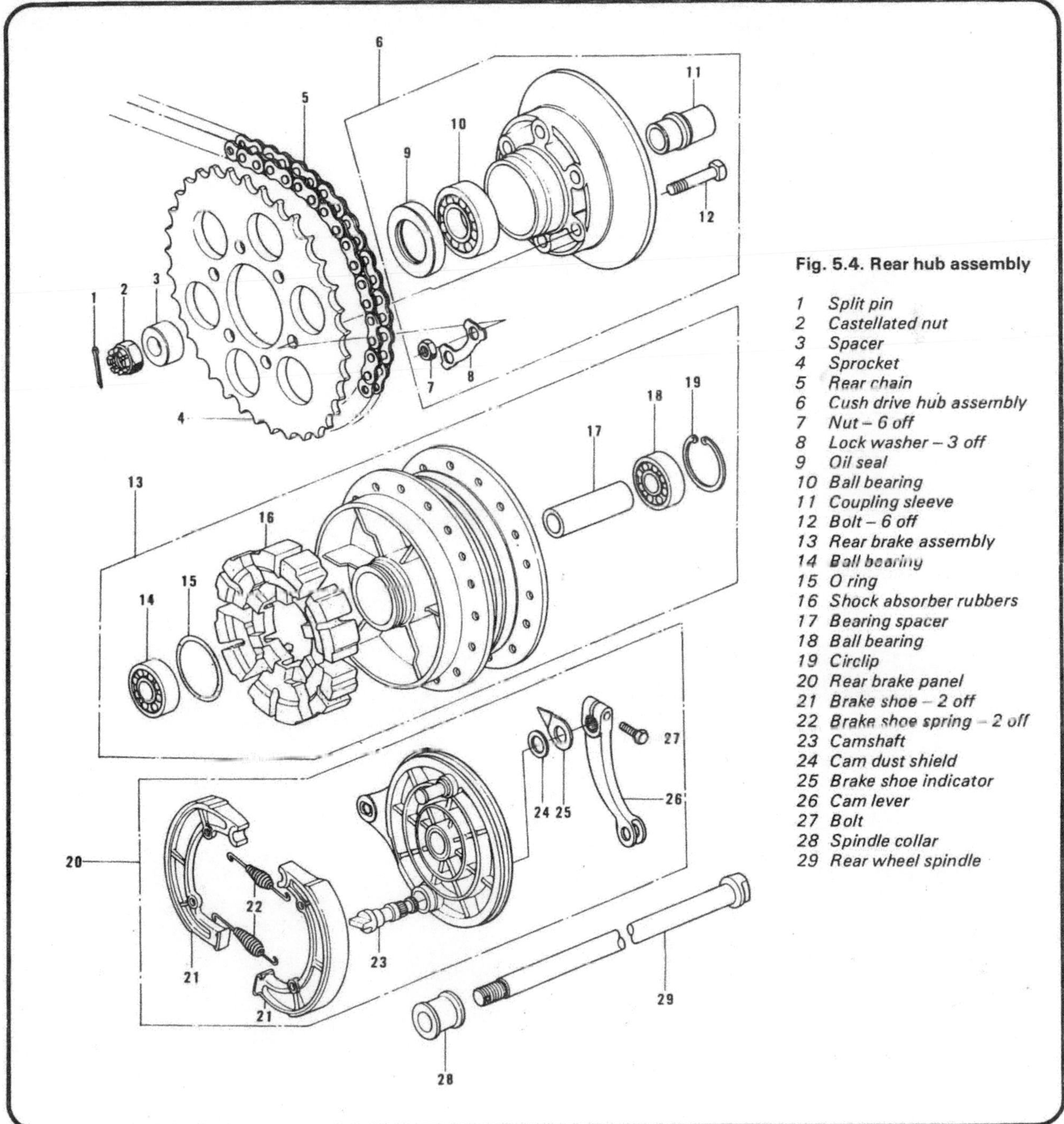

Fig. 5.4. Rear hub assembly

1 *Split pin*
2 *Castellated nut*
3 *Spacer*
4 *Sprocket*
5 *Rear chain*
6 *Cush drive hub assembly*
7 *Nut – 6 off*
8 *Lock washer – 3 off*
9 *Oil seal*
10 *Ball bearing*
11 *Coupling sleeve*
12 *Bolt – 6 off*
13 *Rear brake assembly*
14 *Ball bearing*
15 *O ring*
16 *Shock absorber rubbers*
17 *Bearing spacer*
18 *Ball bearing*
19 *Circlip*
20 *Rear brake panel*
21 *Brake shoe – 2 off*
22 *Brake shoe spring – 2 off*
23 *Camshaft*
24 *Cam dust shield*
25 *Brake shoe indicator*
26 *Cam lever*
27 *Bolt*
28 *Spindle collar*
29 *Rear wheel spindle*

9.2a Remove outer collar, if still in position ...

9.2b ... and withdraw distance piece from bearing

9.2c Oil seal and bearing may now be removed

9.4a A large diameter socket makes a useful drift

9.4b Outer hub bearing and ...

9.4c ... inner hub bearing can be treated similarly

Tyre changing sequence - tubed tyres

Deflate tyre. After pushing tyre beads away from rim flanges push tyre bead into well of rim at point opposite valve. Insert tyre lever adjacent to valve and work bead over edge of rim.

Use two levers to work bead over edge of rim. Note use of rim protectors

Remove inner tube from tyre

When first bead is clear, remove tyre as shown

When fitting, partially inflate inner tube and insert in tyre

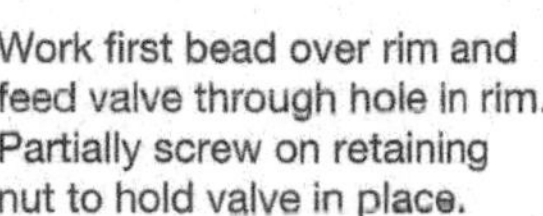

Work first bead over rim and feed valve through hole in rim. Partially screw on retaining nut to hold valve in place.

Check that inner tube is positioned correctly and work second bead over rim using tyre levers. Start at a point opposite valve.

Work final area of bead over rim whilst pushing valve inwards to ensure that inner tube is not trapped

10 Rear brake assembly: examination, renovation and reassembly

1 The rear brake is of the internal expanding variety. Access to the brake shoes is obtained by first removing the rear wheel, and taking off the brake plate to which the shoes are attached.
2 Use a punch to mark the original position of the brake cam and the brake operating lever. Remove the pinch bolt and lever. Remove the dust seal, and the brake shoes, by prying them up evenly and removing them, along with the brake cam.
3 Take off the brake shoe return springs. Inspect the brake drum for a scored or warped condition. If the drum is scored or warped slightly, it is possible to have it turned down on a lathe by a specialist repairer but if the scoring is too deep or the warpage too great, a new replacement is necessary.
4 Inspect the brake shoes for excessive uneven wear, or for oil or grease on the linings. If the impregnation is too bad the shoes will have to be replaced with new ones.

The standard measurement for the brake linings is as follows:

Standard thickness	*Service limit*
4·85 – 5·80 mm (0·191 – 0·228 in)	*2·5 mm (0·098 in)*

5 Inspect the brake return springs for a worn, pitted or collapsed condition, and replace them as necessary.

The brake return spring free length specifications are as follows:

Standard length	*Service limit*
66·0 – 67·0 mm (2·60 – 2·64 in)	*69·0 mm (2·72 in)*

6 Inspect the brake cam and brake plate for signs of wear or damage, and replace as necessary. It cannot be overstressed that wear on these parts is critical if full braking efficiency is to be maintained.
7 Assembly is in the reverse order of dismantling. Use new locking tabs and split pins whenever possible, also smear a light touch of grease on the brake cam and pivot pins during assembly, taking care not to get any grease on the brake linings.

11 Rear sprocket assembly: examination, renovation and replacement

1 The rear wheel sprocket is held to the wheel by six nuts and three locktabs. To remove the sprocket, bend back the locktabs and undo the nuts. The sprocket needs to be renewed only if the teeth are worn, hooked or chipped. It is always good policy to change both sprockets at the same time, also the chain, otherwise very rapid wear will develop.
2 It is not advisable to alter the rear wheel sprocket size or the gearbox sprocket size. The ratios selected by the manufacturer are the ones that give optimum performance with the existing engine power output.

12 Rear cush drive: examination and renovation

1 The cush drive assembly is contained in the left-hand side of the wheel hub. It takes the form of six triangular rubber pads incorporating slots. These engage with vanes on the coupling which is bolted to the rear sprocket. The rubbers engage with ribs on the hub and the whole assembly forms a shock absorber which permits the sprocket to move within certain limits. This cushions any surge or roughness in the transmission which would otherwise convey an impression of harshness.
2 The usual sign that shock absorber rubbers are worn is excessive movement in the sprocket, or rubber dust appearing in between the sprocket and hub. The rubbers should then be taken out and renewed.

13 Rear brake assembly: adjusting

1 If the adjustment of the rear brake is correct, the brake pedal will have a travel of 20 mm to 30 mm (0·8 to 1·2 inch). Adjustment is carried out at the end of the operating rod by an adjusting nut.
2 It may be necessary to change the height of the stop lamp switch if the pedal travel has been altered to any marked extent. Raise the switch for the stop lamp to operate earlier by turning the adjustment nut clockwise.

14 Final drive chain: examination and lubrication

1 As the final drive chain is fully exposed on all models it requires lubrication and adjustment at regular intervals. To adjust the chain, take out the split pin from the rear wheel spindle and slacken the spindle nut. Undo the torque arm bolt, and leave the bolt in position, slacken the chain adjuster locknuts and turn the adjusters inwards to tighten the chain, or outwards to slacken the chain.
2 Chain tension is correct if there is (25 to 40 mm) about $1\frac{1}{4}$ inch slack measured at the centre of the bottom run of the chain between the two sprockets.
3 Do not run the chain too tight to try to compensate for wear, or it will absorb a surprising amount of engine power. Also it can damage the gearbox and rear wheel bearings.
4 All models are equipped with endless chains, having 'O' rings at the end of each pin to retain the grease used during assembly It is not possible to remove the chain for full lubrication in Linklyfe or Chainguard as with normal chains. Lubrication is therefore restricted to frequent cleaning and lubrication with aerosol chain lubricant.
5 Chain wear can be assessed by stretching the chain taut by means of the chain adjusters, and measuring a 20 link section. The measurement should be made between 21 pin centres. This length should normally be 317·5 mm (12·5 in). If worn to 323 mm (12·72 in) or more, the chain must be renewed.
6 Removal of the rear chain for renewal necessitates the removal of the rear wheel and swinging arm assembly, as described in Chapter 4, Section 8. Once the swinging arm has been removed the chain may be lifted off the gearbox sprocket. When fitting a new chain, ensure that the gearbox and rear wheel sprockets are in good condition.

13.1a Adjust rear brake by way of nut on brake rod

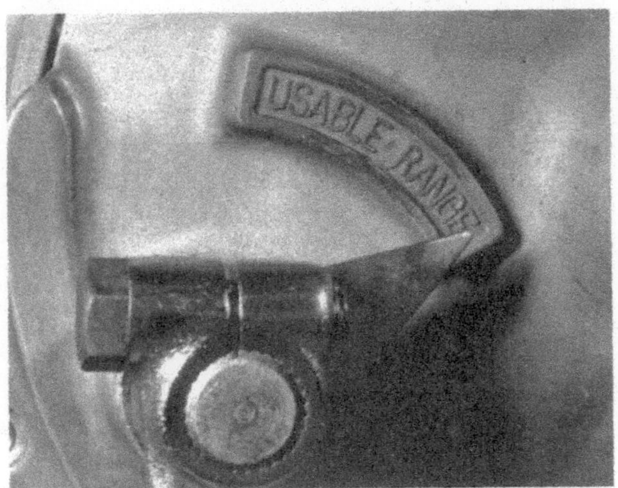

13.1b Pointer indicates range of brake lining wear

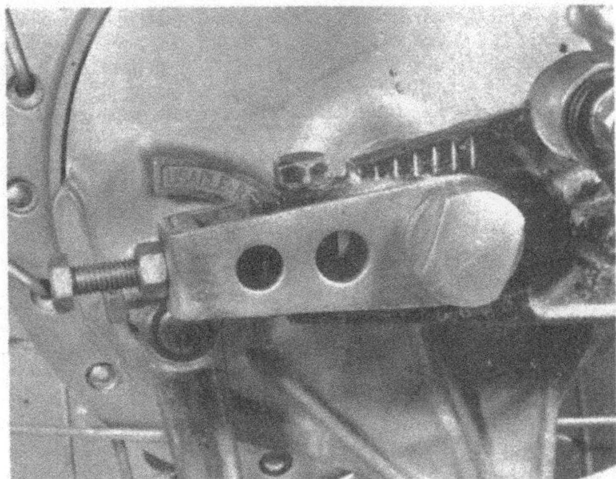
14.1 Keep scribed lines parallel during chain adjustment

14.4 Chain must be kept clean and well lubricated

15 Tyres: removal and replacement

1 At some time or other the need will arise to remove and replace the tyres, either as a result of a puncture or because replacements are necessary to offset wear. To the inexperienced, tyre changing represents a formidable task, yet if a few simple rules are observed and the technique learned, the whole operation is surprisingly simple

2 To remove the tyre from either wheel, first detach the wheel from the machine. Deflate the tyre by removing the valve core, and when the tyre is fully deflated, push the bead from the tyre away from the wheel rim on both sides so that the bead enters the centre well of the rim. Remove the locking ring and push the tyre valve into the tyre itself.

3 Insert a tyre lever close to the valve and lever the edge of the tyre over the outside of the rim. Very little force should be necessary; if resistance is encountered it is probably due to the fact that the tyre beads have not entered the well of the rim, all the way round.

4 Once the tyre has been edged over the wheel rim, it is easy to work round the wheel rim, so that the tyre is completely free from one side. At this stage the inner tube can be removed.

5 Now working from the other side of the wheel, ease the other edge of the tyre over the outside of the wheel rim that is furthest away. Continue to work around the rim until the tyre is completely free from the rim.

6 If a puncture has necessitated the removal of the tyre, reinflate the inner tube and immerse it in a bowl of water to trace the source of the leak. Mark the position of the leak, and deflate the tube. Dry the tube, and clean the area around the puncture with a petrol soaked rag. When the surface has dried, apply rubber solution and allow this to dry before removing the backing from the patch, and applying the patch to the surface.

7 It is best to use a patch of self vulcanizing type, which will form a very permanent repair. Note that it may be necessary to remove a protective covering from the top surface of the patch after it has sealed into position. Inner tubes made from a special synthetic rubber may require a special type of patch and adhesive, if a satisfactory bond is to be achieved.

8 Before replacing the tyre, check the inside to make sure that the article that caused the puncture is not still trapped inside the tyre. Check the outside of the tyre, particularly the tread area to make sure nothing is trapped that may cause a further puncture.

9 If the inner tube has been patched on a number of past occasions, or if there is a tear or large hole, it is preferable to discard it and fit a replacement. Sudden deflation may cause an accident, particularly if it occurs with the front wheel.

10 To replace the tyre, inflate the inner tube for it just to assume a circular shape but only to that amount, and then push the tube into the tyre so that it is enclosed completely. Lay the tyre on the wheel at an angle, and insert the valve through the rim tape and the hole in the wheel rim. Attach the locking ring on the first few threads, sufficient to hold the valve captive in its correct location.

11 Starting at the point furthest from the valve, push the tyre bead over the edge of the wheel rim until it is located in the central well. Continue to work around the tyre in this fashion until the whole of one side of the tyre is on the rim. It may be necessary to use a tyre lever during the final stages.

12 Make sure there is no pull on the tyre valve and again commencing with the area furthest from the valve, ease the other bead of the tyre over the edge of the rim. Finish with the area close to the valve, pushing the valve up into the tyre until the locking ring touches the rim. This will ensure that the inner tube is not trapped when the last section of bead is edged over the rim with a tyre lever.

13 Check that the inner tube is not trapped at any point. Re-inflate the inner tube, and check that the tyre is seating correctly around the wheel rim. There should be a thin rib moulded

around the wall of the tyre on both sides, which should be an equal distance from the wheel rim at all points. If the tyre is unevenly located on the rim, try bouncing the wheel when the tyre is at the recommended pressure. It is probable that one of the beads has not pulled clear of the centre well.

14 Always run the tyres at the recommended pressures and never under or over inflate. The correct pressures for solo use are given in the Specifications Section of this Chapter.

15 Tyre replacement is aided by dusting the side walls, particularly in the vicinity of the beads, with a liberal coating of french chalk. Washing up liquid can also be used to good effect, but this has the disadvantage of causing the inner surface of the wheel rim to rust.

16 Never replace the inner tube and tyre without the rim tape in position. If this precaution is overlooked there is a good chance of the ends of the spoke nipples chafing the inner tube and causing a crop of punctures.

17 Never fit a tyre that has a damaged tread or sidewalls. Apart from legal aspects, there is a very great risk of a blowout, which can have very serious consequences on a two wheeled vehicle.

18 Tyre valves rarely give trouble, but it is always advisable to check whether the valve itself is leaking before removing the tyre. Do not forget to fit the dust cap, which forms an effective extra seal..

16 Tyres: valves and dustcaps

1 Inspect the valves in the inner tubes from time to time making sure that the seal and spring are making an effective seal. There are tyre valve tools available for clearing damaged threads in the valve body, and incorporating thread cleaning for the outside thread of the body. A key is also incorporated for tightening the valve core.

2 The valve caps prevent dirt and foreign matter from entering the valve, and also form an effective second seal so that in the event of the tyre valve sticking, air will not be lost.

3 Note that when a dust cap is fitted for the first time to a balanced wheel, the wheel may have to be rebalanced.

17 Fault diagnosis

Symptom	Cause	Remedy
Handlebars oscillate at low speed	Buckle or flat in wheel rim, most probably front wheel	Check rim alignment by spinning wheel. Correct by retensioning spokes or re-building on new rim.
	Tyre not straight on rim	Check tyre alignment.
Machine lacks power and accelerates poorly	Rear brake binding	Warm brake drum provides best evidence. Re-adjust brake.
Rear brake grabs when applied gently	Ends of brake shoes not chamfered	Chamfer with file.
	Elliptical brake drum	Lightly skim in lathe (specialist attention required).
Front brake feels spongy	Air in hydraulic system	Bleed brake.
Brake pull-off sluggish	Brake cam binding in housing	Free and grease.
	Weak brake shoe springs	Renew if springs have not become displaced.
	Sticking pistons in brake caliper	Overhaul caliper unit.
Harsh transmission	Worn or badly adjusted final drive chain	Adjust or renew as necessary.
	Hooked or badly worn sprockets	Renew as a pair.
	Worn or deteriorating cush drive rubbers	Renew rubbers.

Chapter 6 Electrical system

Contents

Specifications

Battery

Make	Yuasa
Type	YB 10 L
Voltage	12 volts
Capacity	10 Ah
Earth	Negative

Alternator

Make	Nippon Denso
Model	ACQ105
Type	3-phase

Starter motor

Make	Mitsuba
Type	SM-224D

Headlamp

UK and Europe	12v 45/40w bulb type
US	12v 50/35w sealed beam

Bulbs

Stop/Tail	UK 12v 5/21w, US 12v 8/27w
All pilot/warning lamps	12v 3.4w
Indicators	12v 21w (US: 12v 23w)
Front parking lamp	12v 4w (UK only)

Horn

Horn	12v 2.5 amp

1 General description

The Kawasaki Z650 models are equipped with a 12 volt negative (-) earth electrical system. The system comprises a crankshaft-mounted ac (alternating current) generator of the 3-phrase type, the output of which is governed by an electro-mechanical three point regulator to match the electrical demand. A silicon rectifier is incorporated in the circuit to convert the current to dc (direct current) so it can be used to charge the battery.

2 Crankshaft alternator: checking the output

1 The alternator generates all the current required by the machine's electrical circuits; the output is three phase alternating current (ac). The output is changed to direct current (dc) by the rectifier, the voltage being controlled by the voltage regulator. The alternator consists of a rotor and armature. Permanent magnets supply the magnetic field of the rotor, so that no slip rings or brushes are necessary. This makes the rotor practically maintenance free. The armature consists of three

sets of coils wound on laminated steel cores. The coils are connected in a Y pattern, so that there is always a smooth, ample supply of current available.

2 To check the output of the alternator, the battery and rectifier must first be tested so that they are known to be good. If the battery shows less than the required 12 volts it should be fully charged.

3 Remove the left-hand side panel, and disconnect the brown and green leads from the regulator. Connect the two leads together, ensuring that they cannot short against the frame by wrapping them with insulating tape. Unlatch the seat and disconnect the negative (-) battery lead. Obtain a 0-10 amp ammeter, and connect the negative (-) terminal of the ammeter to the negative (-) battery lead, and the positive (+) terminal to the battery negative terminal (-). This, in effect, means that the charging current must pass through the ammeter.

4 To prevent damage to the ammeter, make up a bypass lead with a small crocodile clip at each end, and connect this as a bridge between the ammeter terminals. Obtain a 0.30 volt dc voltmeter, and connect the positive (+) terminal to the positive (+) battery terminal, and the negative (-) terminal to the negative (-) battery terminal.

5 Use the kickstart to start the engine (**not** the electric starter, as this will overload the meters). Set the headlamp on main beam, and switch it on. Disconnect one end of the bypass lead, and note the readings of the two meters at 4000 rpm. These should be in the region of 15 volts and 3 amps. If the readings differ greatly from these figures, it is likely that the alternator is defective. This can be confirmed by taking armature and field coil resistance measurements as described in Section 3.

3 Alternator : checking the armature and field coils

1 If the alternator is suspect, the armature and field coil resistances should be checked, using a multimeter set in the resistance, or ohms, setting.

2 Disconnect the 4-pin alternator output lead connector, and measure the resistance between each pair of the three yellow leads; a total of three readings. These must be between 0.4 - 0.6 ohms. If greatly different from this, remove the unit and take it to a Kawasaki Service Agent for verification and renewal. Similar action should be taken if a measurement between each of the yellow leads in turn, and earth, shows anything other than perfect insulation.

3 Disconnect the green lead from the alternator field coil, and test the resistance between it and earth. The reading must be between 2.7 and 3.4 ohms, a lower reading indicating a short circuit, and a higher reading denoting an open circuit.

4 Silicon rectifier: location and replacement

1 The silicon rectifier fitted to the electrical system converts the ac current produced by the alternator to dc so that it can be used to charge the battery.

2 The rectifier is mounted on the underside of the battery carrier, and is retained by a single mounting nut.

3 The rectifier is a component that cannot be repaired, and if found faulty it has to be replaced with a new unit. Damage to the unit can be caused by running the machine without a battery or if the battery leads are accidentally reversed.

4 The six-diode arrangement (two diodes for each of the dynamo's three output phases) is used to convert ac current into dc current for battery charging, ignition, lighting, and horn circuits. The diodes in the rectifier can only conduct current from negative to positive, and therefore they convert ac to dc. If the rectifier or diodes become faulty they will conduct current either in both directions, or not at all, and therefore lead eventually to a discharged battery.

5 The rectifier can be tested with an ohmmeter. First disconnect the white rectifier plug from the connector panel, and the white lead going to the battery. With the tester set on the R x 10 and R x 100 range, check the resistance between the white rectifier lead and each of the yellow leads, the yellow leads and the white lead, the black lead and each of the yellow leads, and also each yellow and black lead. This involves a total of twelve measurements. The resistance should be low in one direction, and about ten times as great in the other direction. If the readings are the same in either direction for any pair of wires, the rectifier is faulty and should be replaced. The lower reading should be within the $\frac{1}{3}$ scale of zero ohms, regardless of the type of tester used.

6 **Note**: When removing or installing the rectifier, do not loosen or try to tighten the main assembly nut of the rectifier, as this is part of the assembly and should not be disturbed. If disturbed, damage can be caused to the whole rectifier assembly and render it useless. When fitting a new replacement take great care not to disturb the coating over the electrodes, which may peel or flake and destroy the working action.

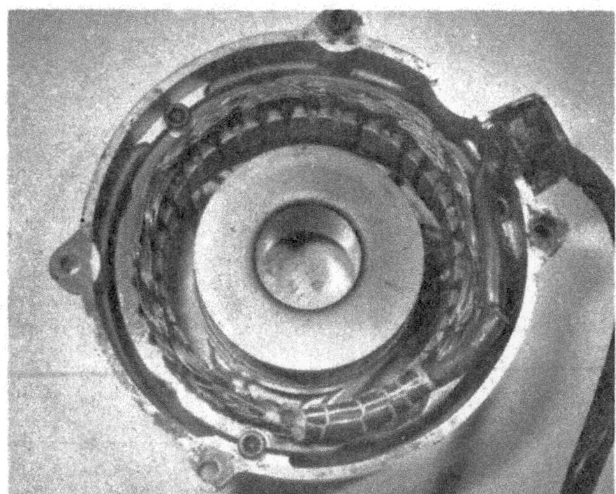

3.2a Stator and field coil winding are housed in outer cover

3.2b Rotor is keyed to tapered mainshaft end

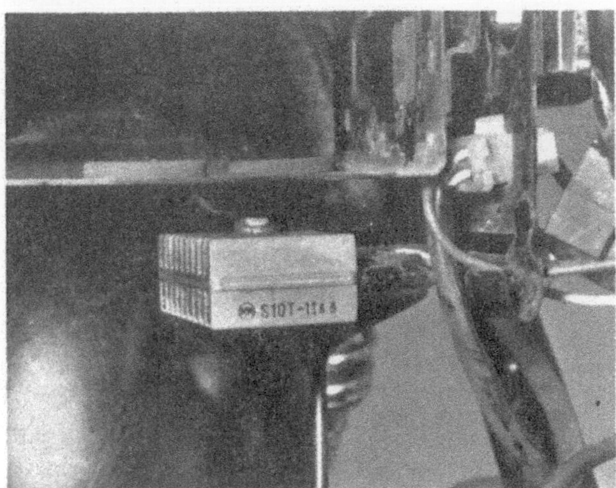
4.2 Rectifier unit is retained to underside of battery tray

5 Voltage regulator: operating principle and testing

1 The voltage regulator fitted to the Z650 model is of the electro-mechanical type. Its function is to handle the power output from the three phase generator, and to limit the voltage to 15-16 volts. It is constructed to control each of the three phases of the alternator output.
2 Two symptoms which would indicate the possibility of a faulty regulator are repeated battery discharging or battery overcharging. A battery overcharged is indicated by the need to top up the electrolyte more frequently than is normal, and also by blowing bulbs in the lighting system when running at high rpms.
3 Discharging of the battery more excessively than is normal is indicated by a battery that when checked reads correctly, but goes dead quickly after being fully charged.
4 If suspected of malfunctioning, the regulator operation may be checked using a voltmeter or multimeter set to 30v dc. Remove the headlamp unit, and separate the 9-pin connector block, thus isolating the electrical system and removing the load from the alternator. Ensure that all the lights etc, are switched off.
5 Connect the voltmeter to the battery, positive to positive (+) to (+), and negative to negative (-) to (-). Start the engine, and gradually bring the speed up to 1600 rpm, at which point the meter should indicate 14-15 volts. **Without backing the throttle off at any point,** gradually bring the engine speed up to 4000 rpm. (On no account let the speed drop slightly, and then rise again, as this will affect the reading. The speed must be increased smoothly from tickover, and the two readings noted. If the throttle is released inadvertently after the first reading, go back to tickover and start the test again). At the second reading point of 4000 rpm, the voltage indicated should still be 14-15 volts, unless the regulator is defective.
6 If the regulator is suspect, it is recommended that the machine be taken to a Kawasaki Service Agent, as any attempt to test or adjust the unit will invalidate the warranty. An authorised Agent will have the equipment and skill necessary to test the unit and effect a repair or adjustment.

6 Battery: examination and maintenance

1 The Kawasaki Z650 models are equipped with a Yuasa 12 volt 10 amp-hour battery. The battery is retained by a rubber strap in a compartment beneath the dualseat. The battery has a translucent plastic case, making a quick visual check of the electrolyte level possible, the two lines denoting the upper and lower levels.
2 The level of the electrolyte in the battery should never be allowed to fall below the lower level water mark. If it does, it must be topped up with distilled water to the upper level, after the initial fill with sulphuric acid of a specific gravity of 1260 to 1280. Also make sure the vent pipe is routed through the proper channel provided to ensure that it discharges clear of the frame parts.
3 It is seldom practicable to repair a cracked battery case because the acid that is already established in the crack will prevent the formation of an effective seal. A cracked battery should be renewed at once because apart from a deterioration in efficiency there will be a considerable amount of corrosion if the acid continues to leak.
4 The battery should be checked every week, and topped up when necessary to the upper electrolyte level. When the machine is laid up for any length of time, it is always advisable to remove the battery and give it a refresher charge every four or six weeks, using a battery charger. Once a battery has been put into service (filled with acid) it must be kept in use, otherwise the cell plates will sulphate and render it useless.

7 Battery: charging procedure

1 The normal charging rate for batteries of up to 14 amp/hour capacity is $1\frac{1}{2}$ to 2 amps. It is permissible to charge at a more rapid rate in an emergency but this shortens the life of the battery, and should be avoided. Always remove the vent caps when recharging the battery, otherwise the gas created within the battery when charging takes place will explode and burst the case with disastrous consequences.

8 Fuses: location and renewal

1 The electrical system is protected by two 10 amp and one 20 amp fuses, housed in a plastic case located beneath the right-hand side panel. The fuses are included to prevent damage to the electrical components in the event of an overload or short circuit.
2 If the fuse blows it should not be renewed until the cause of the short is found. This will involve checking the electrical circuit to correct the fault. If this rule is not observed, the fuse will almost certainly blow again.
3 When a fuse blows and no spare is available a 'get you home' remedy is to wrap the fuse in silver paper before replacing it in the fuse holder. The silver paper will restore electrical continuity by bridging the broken wire within the fuse. Replace the doctored fuse at the earliest opportunity to restore full circuit protection. Make sure any short circuit is eliminated first.
4 Always carry two spare fuses of the correct ratings.

5.1 A: Flasher relay. B: Starter solenoid. C: Regulator unit

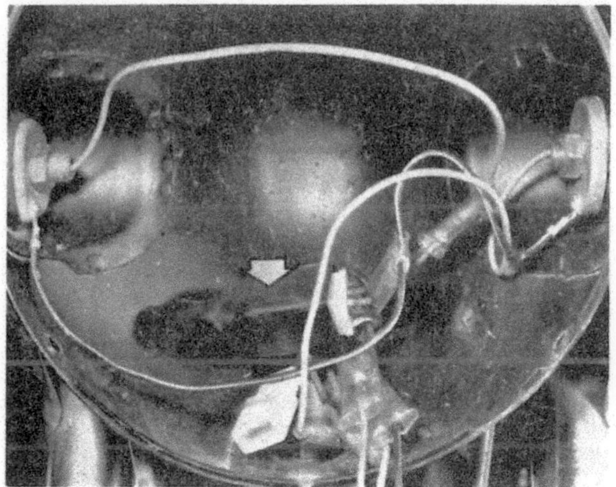
5.4 Separate connector block to isolate lighting circuit

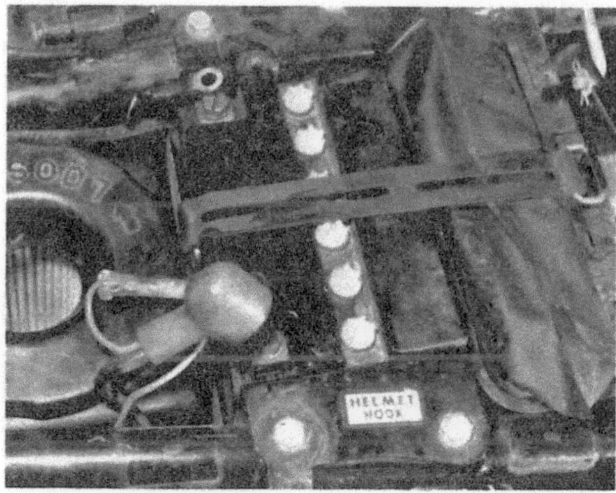

6.1 Battery is housed in compartment beneath the dualseat

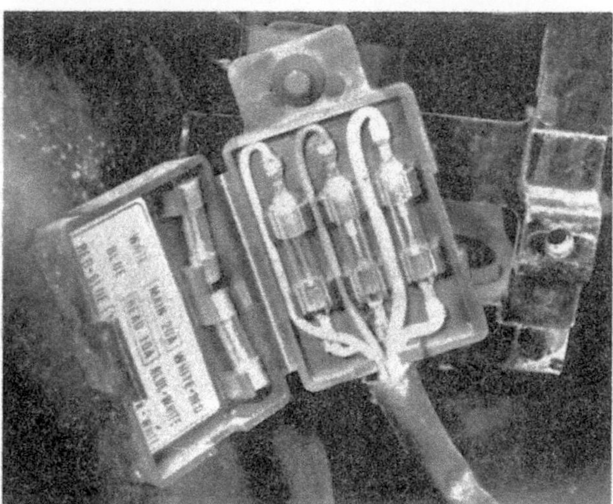
8.1 Fuse box is mounted beneath right-hand side panel

9 Starter motor: removal, examination and replacement

1 An electric starter motor, operated from a small push-button on the right-hand side of the handlebars, provides an alternative and more convenient method of starting the engine, without having to use the kickstart. The starter motor is mounted within a compartment at the rear of the cylinder block, closed by a rectangular chromium plated cover. Current is supplied from the battery via a heavy duty solenoid switch and a cable capable of carrying the very high current demanded by the starter motor on the initial start-up.

2 The starter motor drives a free-wheel type clutch, which is incorporated in the secondary shaft assembly. Maintenance of the clutch assembly is covered in Chapter 1, Section 21. The clutch ensures the starter motor drive is disconnected from the primary transmission immediately the engine starts. It operates on the centrifugal principle; spring loaded rollers take up the drive until the centrifugal force of the rotating engine overcomes their resistance and the drive is automatically disconnected.

3 To remove the starter motor from the engine unit, first disconnect the positive lead from the battery, then the starter motor cable from the solenoid switch. Remove two bolts in the chromium plated cover over the starter motor housing and lift away the cover, complete with gasket. The starter motor is secured to the crankcase by two bolts which pass through the left-hand end of the motor casing. When these bolts are withdrawn, the motor can be prised out of position and lifted out of its compartment, with the heavy duty cable still attached. Once the starter motor has been removed, the starter motor cable may be detached.

4 The parts of the starter motor most likely to require attention are the brushes. The end cover is retained by two long screws which pass through the lugs cast on both end pieces. If the screws are withdrawn, the end cover can be lifted away and the brush gear exposed.

5 Lift up the spring clips which bear on the end of each brush and remove the brushes from their holders. Each brush should have a length of 12.0 - 13.0 mm (0.47 - 0.51 in). The minimum allowable brush length is 6 mm (0.24in). If the brush is shorted it must be renewed.

6 Before the brushes are replaced, make sure that the commutator is clean. The commutator is the copper segments on which the brushes bear. Clean the commutator with a strip of glass paper. Never use emery cloth or 'wet-or-dry' paper as the small abrasive fragments may embed themselves in the soft brass of the commutator and cause excessive wear of the brushes. Finish off the commutator with metal polish to give a smooth surface and finally wipe the segments over with a methylated spirit soaked rag to ensure a grease free surface. Check that the mica insulators, which lie between the segments of the commutator, are undercut. The standard groove depth is 0.5 - 0.8 mm (0.02 - 0.03 in), but if the average groove depth is less than 0.2 mm (0.008 in) the armature should be renewed or returned to a Kawasaki dealer for re-cutting.

7 Replace the brushes in their holders and check that they slide quite freely. Make sure the brushes are replaced in their original positions because they will have worn to the profile of the commutator. Replace and tighten the end cover, then replace the starter motor and cable in the housing, tighten down and remake the electrical connection to the solenoid switch. Check that the starter motor functions correctly before replacing the compartment cover and sealing gasket.

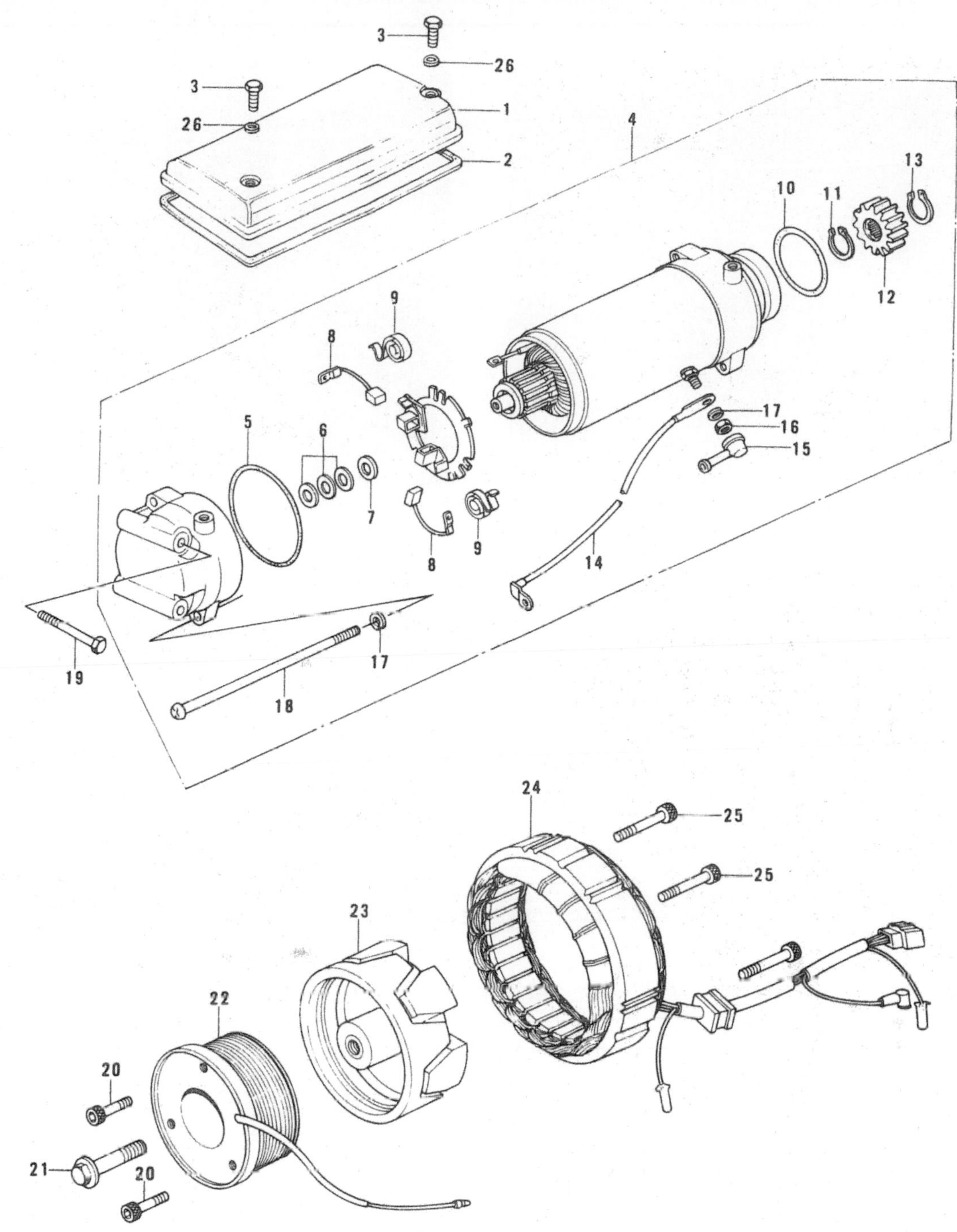

Fig. 6.1 Starter motor – component parts

1 Starter motor cover
2 Starter motor gasket
3 Bolt – 2 off
4 Starter motor assembly
5 O ring
6 Shim – as required
7 Thrust washer
8 Carbon brush – 2 off
9 Brush spring – 2 off
10 O ring
11 Circlip
12 Pinion
13 Circlip
14 Starter motor cable
15 Rubber boot
16 Nut – 2 off
17 Spring washer – 4 off
18 Screw – 2 off
19 Bolt – 2 off
20 Allen bolt – 3 off
21 Bolt
22 Field coil
23 Rotor
24 Stator coil unit
25 Allen bolt – 3 off
26 Washer – 2 off

9.3 Starter motor is retained by two mounting bolts

9.4 Displace spring to release brushes from holder

9.6 Check and clean the commutator surface

9.7 It is easier to reconnect the cable prior to installation

10 Starter solenoid switch: function and location

1 The starter motor switch is designed to work on the electro-magnetic principle. When the starter motor button is depressed, current from the battery passes through windings in the switch solenoid and generates an electro-magnetic force which causes a set of contact points to close. Immediately the points close, the starter motor is energised and a very heavy current is drawn from the battery.

2 This arrangement is used for at least two reasons. Firstly, the starter motor current is drawn only when the button is depressed and is cut off again when pressure on the button is released. This ensures minimum drainage on the battery. Secondly, if the battery is in a low state of charge, there will not be sufficient current to cause the solenoid contacts to close. In consequence, it is not possible to place an excessive drain on the battery which, in some circumstances, can cause the plates to overheat and shed their coatings. If the starter will not operate, first suspect a discharged battery. This can be checked by trying the horn or switching on the lights. If this check shows the battery to be in good shape, suspect the starter switch which should come into action with a pronounced click. It is located behind the left-hand side panel and can be identified by the heavy duty starter cable connected to it. It is not possible to effect a satisfactory repair if the switch malfunctions; it must be renewed.

11 Headlamp: replacing bulbs and adjusting beam height

1 In order to gain access to the headlamp bulbs remove the rim, this is retained by two screws behind the rim. The rim can now be pulled off with the reflector unit complete and the pilot bulb removed.

2 Disconnect the headlamp bulb adaptor from the sealed beam unit and remove the lens retaining ring. The main bulb can now be removed.

3 To adjust headlight beam height, slacken the two turn signal mounting nuts inside the headlamp shell, loosen the mounting bolts underneath the lamp and adjust the vertical aim of the unit to the required position.

4 Adjust the horizontal (left to right) aim of the light by turning the small crosshead screw situated directly in front of the rim. Screwing the screw inwards moves the beam to the right and screwing out moves the beam to the left. On European models the headlamp lens and bulb is of the prefocus type, on USA models the headlamp lens and bulb are a sealed unit, and the whole unit has to be replaced in the event of light failure. Set the beam height with the machine on a level surface 25 feet from a wall so that the centre of the light spot is the same distance as that from the centre of the headlamp to the ground.

10.2 Solenoid unit is retained by rubber mounting block

11.1a Reflector unit is retained by two screws ...

11.1b ... and can be lifted away after these have been released

11.2 UK models have bulb units, US models are sealed beam type

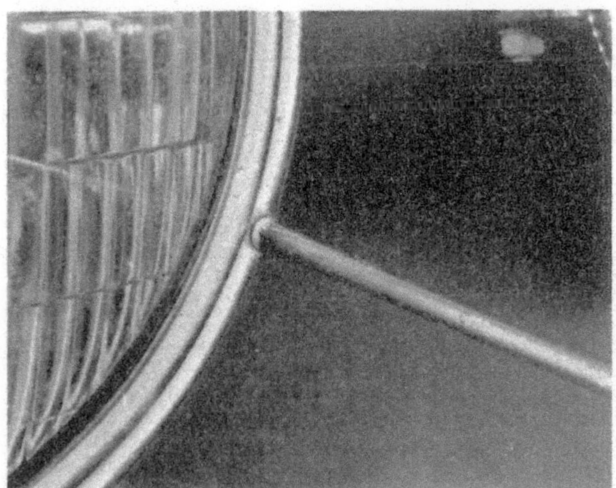

11.4 This screw controls side to side adjustment

12 Stop and tail lamp: replacing the bulb

1 The tail lamp fitted to the Z650 models has a double filament bulb. One lights the rear lamp and the other indicates when the brakes are applied. The brake light is operated by either the front or rear brake lamp switch. The front brake switch is an hydraulic pressure switch, installed in the front hydraulic brake hose. It operates when front brake pressure is applied. The rear brake switch is operated by the rear brake pedal, and is adjustable by altering its position higher or lower in the mounting bracket.

2 Remove the two long screws that retain the rear lamp lens. The bulb can be removed by pushing in and at the same time turning in an anti-clockwise direction. Replace the bulb by reversing the procedure. The bulb has to be renewed if either the tail lamp or brake light filament burns out. When the lens is replaced, make sure the mounting gasket is in good condition and waterproof.

12.2a Indicator and rear lamp lenses are each retained by two screws

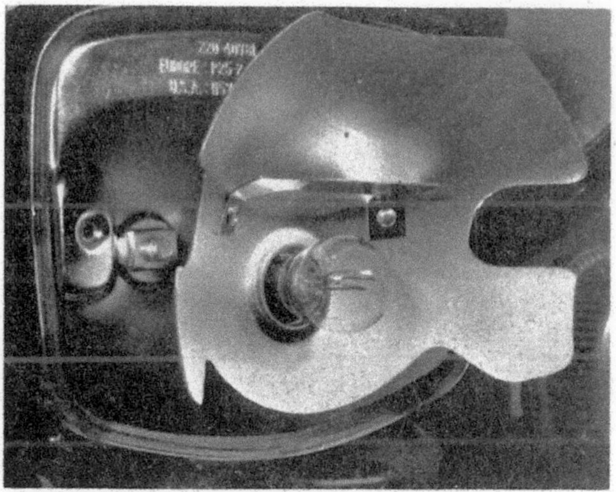
12.2b Stop tail bulb has offset pin bayonet fitting

13 Flashing indicator relay and lamps: location and replacement

1 The flashing indicator relay is fitted to the same electrical panel as the voltage regulator and rectifier, below the dualseat on the right-hand side of the machine. It is mounted in rubber because of the fragile mechanism inside. It is very important not to drop the unit otherwise damage will result.
2 The flashing indicator lamps are fitted to the front and rear of the machine on 'stalks' through which the wires pass. To renew the bulbs remove the two screws that retain each lens, and remove the bayonet type bulbs. These are single filament with a rating of 12 volt 23 watts*. Make sure the rubber gaskets on the base of the lens are in good condition and waterproof, when replacing the lens.
*See Specifications.

14 Speedometer and tachometer: replacing the bulbs

1 The bulbs that fit into the instruments and dash panel are of the small bayonet type, rated at 12 volts, 3.4 watt.
2 The shrouds that cover the instruments have to be removed to expose the bulbs; the bulb holders are a push fit into the back of the panel and are easily removed and replaced.

15 Horn: location and adjustment

1 The horn is adjustable by means of the small screw located at the back of the horn, situated in the top of the front frame gusset. To adjust the volume, turn the screw about half a turn either way until the desired tone is required.
2 If it is necessary to dismantle the horn to clean the contacts, first remove the fuel tank, then remove the horn. Clean the contacts with a fine sand paper and if after this the horn does not work it must be renewed. Make sure the horn is watertight by renewing the gasket when reassembling.

16 Handlebar switches, ignition and lighting switches: examination and replacement

1 The handlebar switches are made up of two halves that clamp together with small crosshead screws; these are situated underneath the switch assemblies. The switches seldom give any trouble, and it is not advisable to take them apart as the parts are so small that difficulties can occur during reassembly, not to mention the time involved. If a switch fails it is far better to fit a new replacement. This is quite a simple task as the wires are fitted into snap connectors.
2 The main ignition switch is located in the centre of the dash panel and is removed by unscrewing the ring nut round the barrel of the switch. The dash panel can then be removed. Take off the front light unit and headlamp shell together with the flasher lamps and remove the switch lower cover and the mounting nut. The ignition switch can now be removed by unplugging the leads.
3 When replacing the ignition switch, the vertical aim of the headlight will have to be readjusted. Also note that the left-hand turn signal wire goes to the green wire, and the right-hand turn signal wire is plugged into the grey wire.

17 Stop lamp switches: adjustment and replacement

1 The rear brake stop lamp switch is located in a bracket above the rear brake pedal and is operated by an expansion spring linked to the rear brake pedal. The body of this switch is threaded to enable it to be raised or lowered.
2 If the rear brake stop lamp is late in operating, slacken the two locknuts and raise the switch body. When the adjustment is correct tighten the locknuts and test. If the stop lamp is early in operation, slacken the locknuts and lower the body in relation to the bracket.
3 As a guide the light should come on when the rear brake pedal has been depressed about 2 cm ($\frac{3}{4}$ in).
4 The hydraulic front brake lamp switch operates the same bulb in the tail light as the rear brake switch. The hydraulic pressure switch operates when the front brake lever is compressed. Adjustment of this switch is not possible. If the pressure switch has to be renewed, the complete switch can be unscrewed after the hydraulic system has been drained from the three way branch pipe situated underneath the headlamp. When a new hydraulic pressure switch is installed it will be necessary to refill and bleed the hydraulic brake system as described in Chapter 5, Section 3.

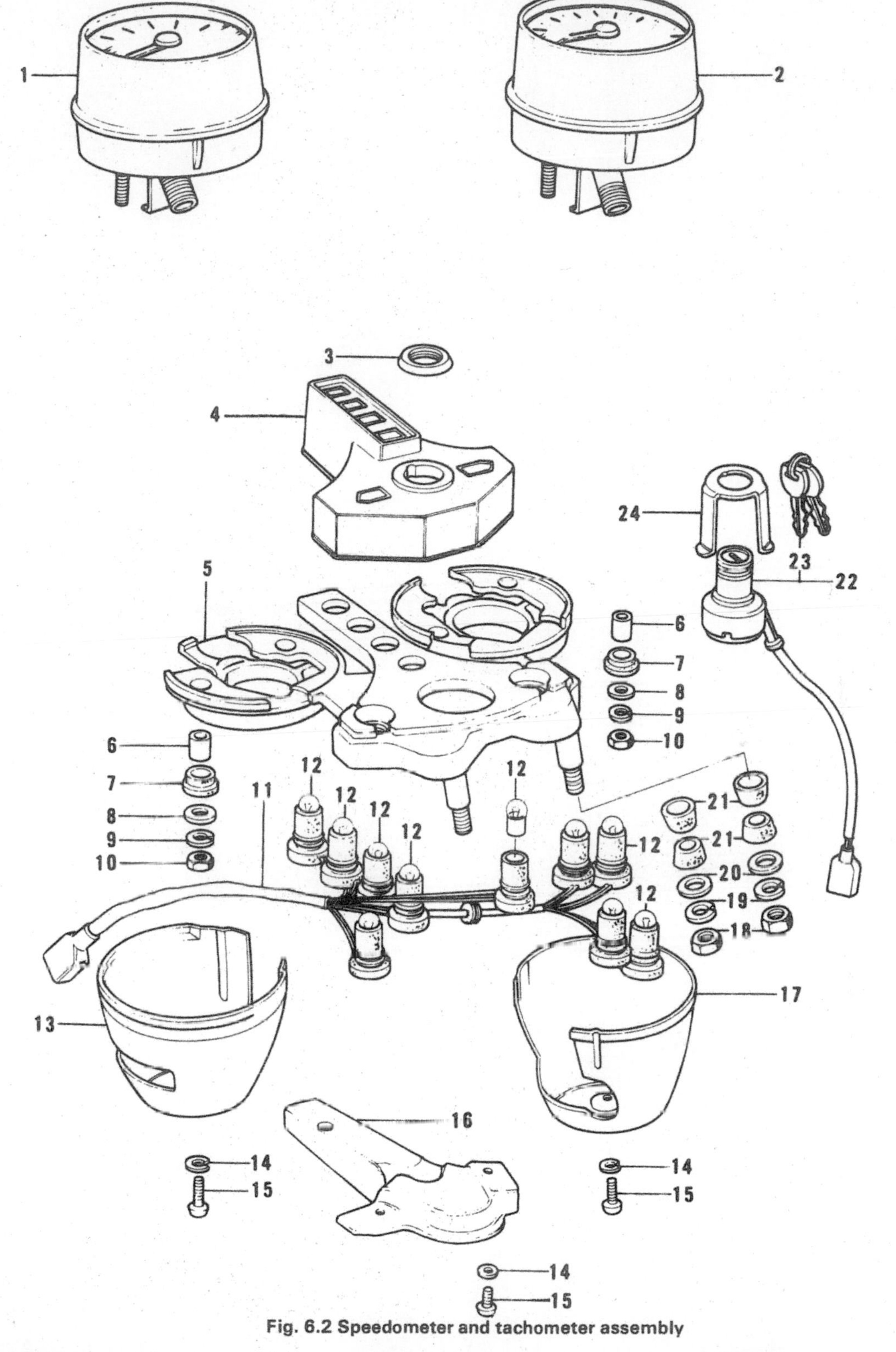

Fig. 6.2 Speedometer and tachometer assembly

1 *Speedometer assembly*
2 *Tachometer*
3 *Lock nut*
4 *Indicator lamp cover*
5 *Bracket*
6 *Collar – 4 off*
7 *Rubber – 4 off*
8 *Plain washer – 4 off*
9 *Spring washer – 4 off*
10 *Nut – 4 off*
11 *Socket assembly*
12 *Bulb – 10 off*
13 *Cover*
14 *Spring washer – 5 off*
15 *Screw – 5 off*
16 *Lower cover*
17 *Tachometer cover*
18 *Nut – 2 off*
19 *Spring washer – 2 off*
20 *Plain washer – 2 off*
21 *Damper – 4 off*
22 *Ignition switch*
23 *Key set*
24 *Ignition switch holder*

13.2 Indicator bulbs are single filament type

14.2a Release instrument shrouds, and remove mounting nuts

14.2b It may prove necessary to remove reservoir cap to gain clearance

14.2c Bulb holders are a push fit in the instrument cases.

15.1 Horn is mounted beneath steering head assembly

16.1a Switch clusters are clamped on either side ...

16.1b ... of the handlebars

17.1 Switch body is adjustable by way of nut and locknut

18 Engine oil pressure switch: removing and replacement

1 The oil pressure switch is mounted inside the contact breaker housing. The oil pressure switch serves to indicate when the oil pressure has dropped due to pump failure, blockage in an oilway or too little oil available to the oil pump. It is not however intended to be used as an indication of correct oil level. If the oil pressure lamp (located in the dash panel) comes on and stays on when the oil is hot and the machine is being rapidly accelerated, the fault is probably the switch, this can sometimes be remedied by revving the engine up past 6000 rpm for a second or two but if this does not put the light out, disconnect the blue wire from the oil pressure switch, and remove the switch. When installing a new switch coat the thread with a sealer to form an oil tight seal. **MAKE SURE IT IS THE SWITCH AT FAULT BEFORE USING THE MACHINE. A GENUINE LUBRICATION PROBLEM WILL CAUSE SEVERE ENGINE DAMAGE.**

18.1 Oil pressure switch is located inside contact breaker housing

19 Fault diagnosis chart on page 124

19 Fault diagnosis: electrical system

Symptom	Cause	Remedy
Complete electrical failure	Blown fuse	Check wiring for loose connections before fitting a new fuse.
	Isolated battery	Check battery connections for signs of corrosion.
Constant blowing of bulbs	Vibration or poor earth connections	Check bulb holders, check earth return connections
Dim lights, horn and starter do not work.	Discharged battery	Recharge battery with a battery charger. Check generator for output.
Starter motor sluggish or will not work	Worn brushes	Remove starter motor and replace with new brushes. Clean commutator
Flashing lights will not flash	Faulty relay unit	Replace with a new relay unit.
	Bad earth.	Check flasher lamp bulb holders for good earth.

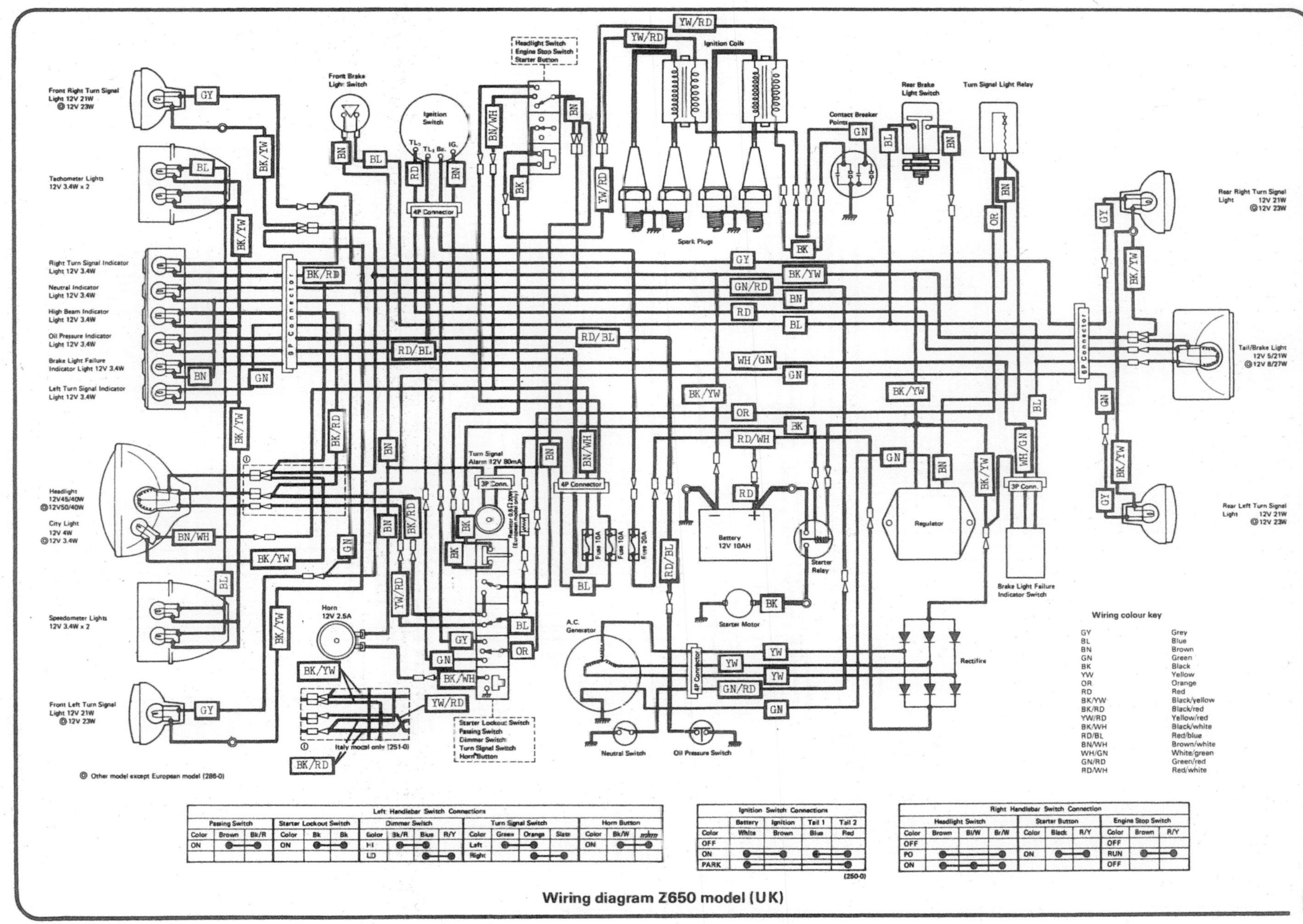

Left Handlebar Switch Connections

Passing Switch			Starter Lockout Switch			Dimmer Switch				Turn Signal Switch				Horn Button		
Color	Brown	Bk/R	Color	Bk	Bk	Color	Bk/R	Blue	R/Y	Color	Green	Orange	Slate	Color	Bk/W	(earth)
ON	●	●	ON	●	●	HI	●	●		Left	●	●		ON	●	●
						LO		●	●	Right		●	●			

Ignition Switch Connections

	Battery	Ignition	Tail 1	Tail 2
Color	White	Brown	Blue	Red
OFF				
ON	●	●	●	●
PARK	●			●

(250-0)

Right Handlebar Switch Connection

Headlight Switch				Starter Button			Engine Stop Switch		
Color	Brown	Bl/W	Br/W	Color	Black	R/Y	Color	Brown	R/Y
OFF							OFF		
PO	●		●	ON	●	●	RUN	●	●
ON	●	●	●				OFF		

Wiring diagram Z650 model (UK)

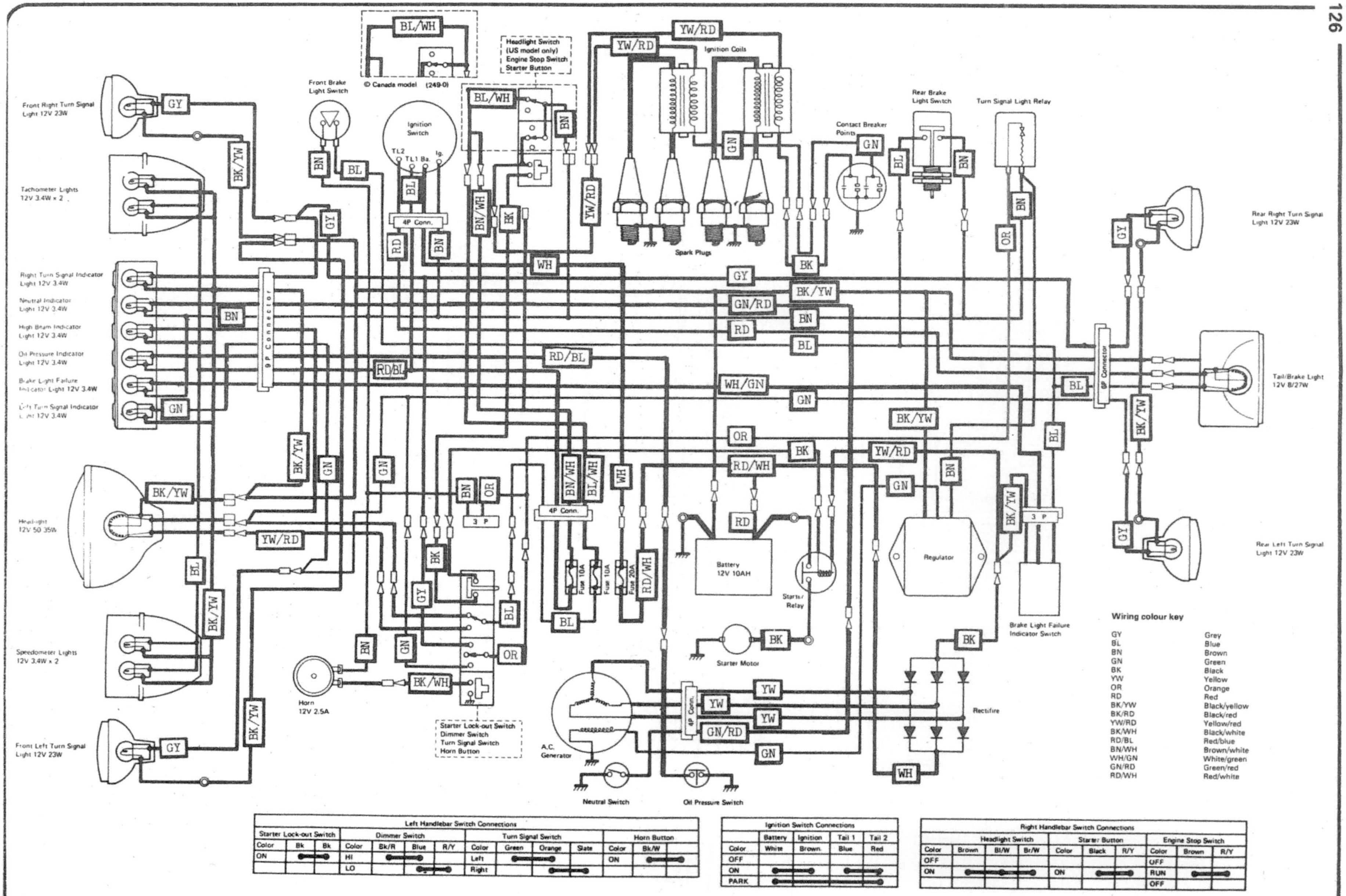

Front Right Turn Signal Light 12V 23W
Tachometer Lights 12V 3.4W x 2
Right Turn Signal Indicator Light 12V 3.4W
Neutral Indicator Light 12V 3.4W
High Beam Indicator Light 12V 3.4W
Oil Pressure Indicator Light 12V 3.4W
Brake Light Failure Indicator Light 12V 3.4W
Left Turn Signal Indicator Light 12V 3.4W
Headlight 12V 50 35W
Speedometer Lights 12V 3.4W x 2
Front Left Turn Signal Light 12V 23W
Front Brake Light Switch
© Canada model (249-0)
Ignition Switch
TL2 TL1 Ba. Ig.
4P Conn.
Headlight Switch (US model only)
Engine Stop Switch
Starter Button
9P Connector
3 P
Horn 12V 2.5A
Starter Lock-out Switch
Dimmer Switch
Turn Signal Switch
Horn Button
Fuse 10A
Fuse 10A
Fuse 20A
A.C. Generator
Neutral Switch
Oil Pressure Switch
Ignition Coils
Spark Plugs
Contact Breaker Points
Rear Brake Light Switch
Turn Signal Light Relay
Battery 12V 10AH
Starter Relay
Starter Motor
Regulator
Rectifire
4P Conn.
Brake Light Failure Indicator Switch
Rear Right Turn Signal Light 12V 23W
Tail/Brake Light 12V 8/27W
Rear Left Turn Signal Light 12V 23W

Wiring colour key

GY	Grey
BL	Blue
BN	Brown
GN	Green
BK	Black
YW	Yellow
OR	Orange
RD	Red
BK/YW	Black/yellow
BK/RD	Black/red
YW/RD	Yellow/red
BK/WH	Black/white
RD/BL	Red/blue
BN/WH	Brown/white
WH/GN	White/green
GN/RD	Green/red
RD/WH	Red/white

Left Handlebar Switch Connections

Starter Lock-out Switch			Dimmer Switch				Turn Signal Switch				Horn Button		
Color	Bk	Bk	Color	Bk/R	Blue	R/Y	Color	Green	Orange	Slate	Color	Bk/W	
ON	●	●	HI	●	●		Left	●	●		ON	●	●
			LO		●	●	Right		●	●			

Ignition Switch Connections

	Battery	Ignition	Tail 1	Tail 2
Color	White	Brown	Blue	Red
OFF				
ON	●	●	●	●
PARK	●			●

Right Handlebar Switch Connections

Headlight Switch				Starter Button			Engine Stop Switch		
Color	Brown	Bl/W	Br/W	Color	Black	R/Y	Color	Brown	R/Y
OFF							OFF		
ON	●	●	●	ON	●	●	RUN	●	●
							OFF		

Wiring diagram KZ650 model (USA)

Metric conversion tables

Inches	Decimals	Millimetres	Millimetres to Inches		Inches to Millimetres	
			mm	Inches	Inches	mm
1/64	0.015625	0.3969	0.01	0.00039	0.001	0.0254
1/32	0.03125	0.7937	0.02	0.00079	0.002	0.0508
3/64	0.046875	1.1906	0.03	0.00118	0.003	0.0762
1/16	0.0625	1.5875	0.04	0.00157	0.004	0.1016
5/64	0.078125	1.9844	0.05	0.00197	0.005	0.1270
3/32	0.09375	2.3812	0.06	0.00236	0.006	0.1524
7/64	0.109375	2.7781	0.07	0.00276	0.007	0.1778
1/8	0.125	3.1750	0.08	0.00315	0.008	0.2032
9/64	0.140625	3.5719	0.09	0.00354	0.009	0.2286
5/32	0.15625	3.9687	0.1	0.00394	0.01	0.254
11/64	0.171875	4.3656	0.2	0.00787	0.02	0.508
3/16	0.1875	4.7625	0.3	0.01181	0.03	0.762
13/64	0.203125	5.1594	0.4	0.01575	0.04	1.016
7/32	0.21875	5.5562	0.5	0.01969	0.05	1.270
15/64	0.234375	5.9531	0.6	0.02362	0.06	1.524
1/4	0.25	6.3500	0.7	0.02756	0.07	1.778
17/64	0.265625	6.7469	0.8	0.03150	0.08	2.032
9/32	0.28125	7.1437	0.9	0.03543	0.09	2.286
19/64	0.296875	7.5406	1	0.03937	0.1	2.54
5/16	0.3125	7.9375	2	0.07874	0.2	5.08
21/64	0.328125	8.3344	3	0.11811	0.3	7.62
11/32	0.34375	8.7312	4	0.15748	0.4	10.16
23/64	0.359375	9.1281	5	0.19685	0.5	12.70
3/8	0.375	9.5250	6	0.23622	0.6	15.24
25/64	0.390625	9.9219	7	0.27559	0.7	17.78
13/32	0.40625	10.3187	8	0.31496	0.8	20.32
27/64	0.421875	10.7156	9	0.35433	0.9	22.86
7/16	0.4375	11.1125	10	0.39370	1	25.4
29/64	0.453125	11.5094	11	0.43307	2	50.8
15/32	0.46875	11.9062	12	0.47244	3	76.2
31/64	0.48375	12.3031	13	0.51181	4	101.6
1/2	0.5	12.7000	14	0.55118	5	127.0
33/64	0.515625	13.0969	15	0.59055	6	152.4
17/32	0.53125	13.4937	16	0.62992	7	177.8
35/64	0.546875	13.8906	17	0.66929	8	203.2
9/16	0.5625	14.2875	18	0.70866	9	228.6
37/64	0.578125	14.6844	19	0.74803	10	254.0
19/32	0.59375	15.0812	20	0.78740	11	279.4
39/64	0.609375	15.4781	21	0.82677	12	304.8
5/8	0.625	15.8750	22	0.86614	13	330.2
41/64	0.640625	16.2719	23	0.90551	14	355.6
21/32	0.65625	16.6687	24	0.94488	15	381.0
43/64	0.671875	17.0656	25	0.98425	16	406.4
11/16	0.6875	17.4625	26	1.02362	17	431.8
45/64	0.703125	17.8594	27	1.06299	18	457.2
23/32	0.71875	18.2562	28	1.10236	19	482.6
47/64	0.734375	18.6531	29	1.14173	20	508.0
3/4	0.75	19.0500	30	1.18110	21	533.4
49/64	0.765625	19.4469	31	1.22047	22	558.8
25/32	0.78125	19.8437	32	1.25984	23	584.2
51/64	0.796875	20.2406	33	1.29921	24	609.6
13/16	0.8125	20.6375	34	1.33858	25	635.0
53/64	0.828125	21.0344	35	1.37795	26	660.4
27/32	0.84375	21.4312	36	1.41732	27	685.8
55/64	0.859375	21.8281	37	1.4567	28	711.2
7/8	0.875	22.2250	38	1.4961	29	736.6
57/64	0.890625	22.6219	39	1.5354	30	762.0
29/32	0.90625	23.0187	40	1.5748	31	787.4
59/64	0.921875	23.4156	41	1.6142	32	812.8
15/16	0.9375	23.8125	42	1.6535	33	838.2
61/64	0.953125	24.2094	43	1.6929	34	863.6
31/32	0.96875	24.6062	44	1.7323	35	889.0
63/64	0.984375	25.0031	45	1.7717	36	914.4

1 Imperial gallon = 8 Imp pints = 1.20 US gallons = 277.42 cu in = 4.54 litres

1 US gallon = 4 US quarts = 0.83 Imp gallon = 231 cu in = 3.78 litres

1 Litre = 0.21 Imp gallon = 0.26 US gallon = 61.02 cu in = 1000 cc

Miles to Kilometres	
1	1.61
2	3.22
3	4.83
4	6.44
5	8.05
6	9.66
7	11.27
8	12.88
9	14.48
10	16.09
20	32.19
30	48.28
40	64.37
50	80.47
60	96.56
70	112.65
80	128.75
90	144.84
100	160.93

Kilometres to Miles	
1	0.62
2	1.24
3	1.86
4	2.49
5	3.11
6	3.73
7	4.35
8	4.97
9	5.59
10	6.21
20	12.43
30	18.64
40	24.85
50	31.07
60	37.28
70	43.50
80	49.71
90	55.92
100	62.14

lbf ft to kgf m	
1	0.138
2	0.276
3	0.414
4	0.553
5	0.691
6	0.829
7	0.967
8	1.106
9	1.244
10	1.382
20	2.765
30	4.147

kgf m to lbf ft	
1	7.233
2	14.466
3	21.699
4	28.932
5	36.165
6	43.398
7	50.631
8	57.864
9	65.097
10	72.330
20	144.660
30	216.990

lbf/in^2 to kgf/cm^2	
1	0.07
2	0.14
3	0.21
4	0.28
5	0.35
6	0.42
7	0.49
8	0.56
9	0.63
10	0.70
20	1.41
30	2.11

kgf/cm^2 to lbf/in^2	
1	14.22
2	28.50
3	42.67
4	56.89
5	71.12
6	85.34
7	99.56
8	113.79
9	128.00
10	142.23
20	284.47
30	426.70

English/American terminology

Because this book has been written in England, British English component names, phrases and spellings have been used throughout. American English usage is quite often different and whereas normally no confusion should occur, a list of equivalent terminology is given below.

English	American	English	American
Air filter	Air cleaner	Number plate	License plate
Alignment (headlamp)	Aim	Output or layshaft	Countershaft
Allen screw/key	Socket screw/wrench	Panniers	Side cases
Anticlockwise	Counterclockwise	Paraffin	Kerosene
Bottom/top gear	Low/high gear	Petrol	Gasoline
Bottom/top yoke	Bottom/top triple clamp	Petrol/fuel tank	Gas tank
Bush	Bushing	Pinking	Pinging
Carburettor	Carburotor	Rear suspension unit	Rear shock absorber
Catch	Latch	Rocker cover	Valve cover
Circlip	Snap ring	Selector	Shifter
Clutch drum	Clutch housing	Self-locking pliers	Vise-grips
Dip switch	Dimmer switch	Side or parking lamp	Parking or auxiliary light
Disulphide	Disulfide	Side or prop stand	Kick stand
Dynamo	DC generator	Silencer	Muffler
Earth	Ground	Spanner	Wrench
End float	End play	Split pin	Cotter pin
Engineer's blue	Machinist's dye	Stanchion	Tube
Exhaust pipe	Header	Sulphuric	Sulfuric
Fault diagnosis	Trouble shooting	Sump	Oil pan
Float chamber	Float bowl	Swinging arm	Swingarm
Footrest	Footpeg	Tab washer	Lock washer
Fuel/petrol tap	Petcock	Top box	Trunk
Gaiter	Boot	Torch	Flashlight
Gearbox	Transmission	Two/four stroke	Two/four cycle
Gearchange	Shift	Tyre	Tire
Gudgeon pin	Wrist/piston pin	Valve collar	Valve retainer
Indicator	Turn signal	Valve collets	Valve cotters
Inlet	Intake	Vice	Vise
Input shaft or mainshaft	Mainshaft	Wheel spindle	Axle
Kickstart	Kickstarter	White spirit	Stoddard solvent
Lower leg	Slider	Windscreen	Windshield
Mudguard	Fender		

Index